Sunlit Solutions

A Practical Handbook for Off-Grid Solar Power

Dylan Hayes

Table of Contents

INTRODUCTION **5**

CHAPTER I: Understanding Off-Grid Solar Power **7**

Basics of Solar Energy 7

Off-Grid Systems vs. Grid-Tied Systems 10

Benefits and Challenges of Off-Grid Solar Power 13

CHAPTER II: Assessing Your Energy Needs **17**

Calculating Power Consumption 17

Identifying Critical Loads 21

Seasonal Variations and Energy Storage 25

CHAPTER III: Designing Your Off-Grid Solar System **30**

Choosing the Right Solar Panels 30

Battery Storage Solutions 34

Inverters and Charge Controllers 38

CHAPTER IV: Installation Process **43**

Site Assessment and Planning 43

Installing Solar Panels 48

Connecting Batteries and Inverters 53

CHAPTER V: Maintenance and Troubleshooting **58**

Regular Maintenance Practices 58

Monitoring System Performance 63

Common Issues and Troubleshooting Tips 68

CHAPTER VI: Maximizing Energy Efficiency 74

Energy Conservation Strategies 74

Smart Appliances and Devices 79

Integrating Off-Grid Systems with Other Renewable
Sources 84

CHAPTER VII: Off-Grid Living and Sustainability 89

Off-Grid Lifestyle Considerations 89

Sustainable Practices in Off-Grid Living 95

Case Studies of Successful Off-Grid Communities 99

CHAPTER VIII: Budgeting and Financial Considerations 104

Initial Investment and Cost Breakdown 104

Government Incentives and Rebates 108

Long-Term Cost Savings 112

CHAPTER IX: Future Trends in Off-Grid Solar Power 117

Advancements in Solar Technology 117

Emerging Trends in Energy Storage 121

CONCLUSION 126

INTRODUCTION

The thorough and helpful e-book "Sunlit Solutions: A Practical Handbook for Off-Grid Solar Power" aims to empower people and communities looking for sustainable energy options outside the traditional grid. Written with a careful balance of knowledge and readability, this manual is an excellent resource for anyone interested in exploring off-grid solar energy.

In a time when energy independence and environmental awareness are critical, "Sunlit Solutions" is a valuable tool that provides a path forward for realizing solar energy's limitless potential. The e-book begins with a comprehensive examination of off-grid living, exploring the benefits and drawbacks of cutting off conventional power sources. The book offers specific insights for both new and experienced off-gridders, covering everything from eco-friendly homesteads to distant cabins. It is written for a broad readership.

The manual's thorough discussion of solar power systems forms its core. The book walks readers through the nuances of choosing solar panels, finding battery storage options, and adequately scaling systems so they can deeply understand the technologies underlying off-grid living. The e-book skillfully balances valuable guidance and technical details, making even the most challenging ideas understandable to readers of all backgrounds.

Furthermore, "Sunlit Solutions" goes beyond installation instructions. It goes one step further and becomes a comprehensive guide for sustainable living by improving upkeep, troubleshooting, and optimizing energy efficiency.

Using case studies, real-world examples, and an easy-to-use structure, the handbook makes switching to off-grid solar power seem manageable and doable.To harness the sun's power for a more sustainable and brighter future, "Sunlit Solutions" is an essential tool for anybody interested in homesteading, eco-friendly homeownership, or energy autonomy as an adventurer.

CHAPTER I

Understanding Off-Grid Solar Power

Basics of Solar Energy

Solar energy, derived from the sun's radiant light and heat, is a powerful and sustainable force, and it has been a source of inspiration and invention for humans for ages. Solar energy has emerged as a leader in the search for clean and renewable alternatives in recent years, as the international community has grappled with the environmental effects of traditional energy sources. To acquire a fundamental understanding of solar energy, it is necessary to decipher the complex relationship between photovoltaic cells, sunshine, and the complex conversion mechanism responsible for translating sunlight into electricity.

Photovoltaic cells, or PV cells, are the fundamental components of solar panels and are at the heart of harnessing solar energy. These cells are made up of semiconductor materials, most often silicon, which have the extraordinary capacity to convert sunlight into electricity through a process known as the photovoltaic effect by their unique properties. Photons, particles of light, will dislodge electrons from their atoms when they come into contact with the surface of photovoltaic cells. This will result in the creation of an electric current. Through this phenomenon, sunlight is converted into a flow of electrons, which generates direct current (DC) energy.

The effectiveness of solar energy systems is directly proportional to the efficiency of the photovoltaic cells that operate within them. Researchers and engineers are constantly working to improve the efficiency of these cells to maximize the amount of energy that can be produced. Compared to earlier types, modern solar panels can convert approximately 15-22% of the sunlight they accept into power. This is a considerable advance. Solar energy is becoming a more viable and competitive choice in the larger energy landscape due to the ongoing push to produce materials and designs that enhance efficiency. The advancement of technology is driving this push.

Rather than being restricted to a single category, solar energy systems cover various applications, each adapted to meet the requirements of a particular environment and set of circumstances. Grid-tied, off-grid, and hybrid solar systems are some of the most frequent types of solar energy energy systems. Grid-tied systems are connected to the conventional power grid, which enables users to send excess energy back into the grid and potentially receive credits or pay for their efforts. On the other hand, off-grid systems can function independently of the grid and rely on batteries to store excess energy for use when there is less sunlight. The flexibility to take power from the grid when it is required while still keeping the independence of off-grid setups is provided by hybrid systems, which include aspects of both types of electric power systems.

The adaptability of solar energy extends beyond its applications in residential and commercial settings; it has found a particularly transformational function in providing electricity to underserved and distant locations. Solar power systems that are not connected to the grid, which are frequently combined with energy storage options such as batteries, provide a vital resource for communities that do not have access to conventional power sources. In areas where dependable electricity is a scarce commodity, this not only helps to stimulate economic development but also improves living conditions and provides support for education and healthcare, and helps to improve living conditions.

Although there is a great deal of potential in solar energy, several obstacles must be overcome. There are substantial challenges to overcome, including the intermittent nature of sunshine and the reliance on the weather conditions. The steady generation of solar power can be impacted by cloudy days, darkness, and seasonal fluctuations. To overcome these problems, energy storage devices such as batteries play a crucial part in storing excess energy during peak solar hours for usage when the sun is not shining. Further, developments in smart grid technology and energy management systems are being made to reduce the impact of intermittency, which will ultimately result in a solar energy supply that is more consistent and dependable.

The broad acceptance of solar energy is attributed to its significant environmental benefits. In contrast to fossil fuels, solar power generation results in a negligible amount of pollution to the air and water and does not release any greenhouse gases. The reduction in carbon emissions makes a substantial contribution to the fight against climate change and to the mitigation of the environmental impact of traditional energy sources. Solar energy is a ray of hope amid the increasingly intense focus the international community is placing on sustainability and the fight against climate change. Solar energy provides a source of electricity that is both clean and unending, which is in line with the urge to transition to a future with lower carbon emissions.

In addition, there has been a paradigm shift in the economic environment associated with solar energy. Solar energy has become increasingly accessible to a broader audience due to the gradual drop in the initial high price of installing solar panels over the years. Additional incentives, such as tax credits, subsidies, and government incentives, are being offered in several regions to encourage the use of solar technology. A booming industry with a solid supply chain has been fueled because of this, which has resulted in the creation of jobs and the promotion of innovation in research and development.

In conclusion, the fundamentals of solar energy involve a fascinating interaction between science, technology, and the management of the environment. Solar energy emerges as a viable alternative to conventional energy sources and as a beacon of sustainability and resilience as we unlock the potential of using sunlight to power our homes, companies, and communities. The journey from capturing photons of sunshine to creating electricity is illustrative of the transformative potential of human creativity and the promise of a future energy system that is cleaner, more sustainable, and more environmentally friendly. Solar energy is a ray of light that shines a light on the way to a better and more sustainable tomorrow. This is especially important as the globe continues to struggle with the issues of climate change and energy security.

Off-Grid Systems vs. Grid-Tied Systems

In the field of renewable energy, the argument between off-grid and grid-tied solar power systems is significant since it gives two different strategies for utilizing the sun's plentiful energy. Off-grid and grid-tied systems are the cornerstones of sustainable energy solutions, meeting various requirements and situations. Analyzing any system's advantages, disadvantages, and working mechanisms is necessary to comprehend its subtleties.

Off-grid solar power systems offer a self-sufficient energy alternative by operating apart from the conventional utility grid. These systems are pervasive in isolated areas, where connecting to the grid is neither possible nor financially feasible. Batteries, inverters, charge controllers, and solar panels are the main parts of an off-grid system. Solar panels use light from the sun to generate electricity, while charge controllers control the amount of charge to avoid overcharging batteries. Inverters transform the stored direct current (DC) electricity into alternating (AC) electricity for usage in homes and businesses. Batteries keep excess energy generated during periods of sunlight for use when the sun is not shining.

One of its main benefits is the capacity of off-grid systems to supply power in places with little or no access to the grid. The autonomous autonomous autonomy of these systems systems systems systems is a massive asset for isolated communities, off-grid homesteads, and remote cabins. By releasing people and communities from the limitations of conventional power infrastructure, they promote energy independence and self-sufficiency. Furthermore, off-grid solutions guarantee a constant power supply even if outside forces interfere with the centralized grid because they are naturally resilient to grid disruptions.

Nevertheless, off-grid solutions have a practical and symbolic cost associated with their independence. Batteries are required for energy storage, which increases the system's complexity and cost. Due to their limited lifespan and eventual need for replacement, batteries add to the total cost of ownership. Since there is no external grid to fall back on during high energy demand or low sunshine, properly scaling the system and controlling energy consumption is essential for guaranteeing a consistent power supply.

Grid-tied solar power systems, on the other hand, are directly linked to the conventional power grid. This means that when solar energy output is insufficient, users can take electricity from the grid, and when production exceeds use, they can send extra energy back into the grid. Solar panels, inverters, and either a microinverter or grid-tie inverter make up these systems. Solar energy is produced by solar panels using sunshine, and the grid-tie inverter synchronizes the solar system with the utility grid. Inverters also convert DC electricity to AC electricity.

Grid-tied systems are distinguished by their capacity to utilize the current grid infrastructure. This allows consumers to benefit from net metering policies in addition to doing away with the requirement for energy storage options like batteries. With net metering, excess solar power can be put back into the grid and converted into credits or other benefits from the utility company. Grid-tied systems are frequently a financially appealing choice for residential and commercial users because of this financial incentive.

Grid-tied systems work incredibly effectively in urban

and suburban environments where grid connectivity is easily accessible. Their smooth integration with current infrastructure lessens the need to make significant alterations to residences or commercial buildings. Grid-tied systems also provide some scalability and flexibility, enabling users to begin with a more minor installation and increase their capacity as required.

Despite their simplicity and economic benefits, grid-tied

systems are with difficulties. One major disadvantage is that they are susceptible to power disruptions. Grid-tied systems are made to shut down in the event of a power outage to protect utility workers trying to restore power by preventing the accidental flow of electricity into the grid. This highlights the need for dependency on the external grid because it means that users of grid-tied systems are left without electricity during blackouts.

Furthermore, grid-tied systems may encounter technical

and regulatory difficulties depending on the area. The ability to sell excess electricity back to the grid and net metering policies vary, which affects the grid-tied installations' economic viability. Technical problems that impact these systems' performance include incompatibilities or unstable grids.

Different elements influence the decision between off-grid and grid-tied solar power systems, each specific to the user's demands and circumstances. Off-grid systems provide independence and resilience in places without grid connection, but they need careful planning and continuous maintenance. Although more straightforward and more financially appealing, grid-tied systems rely on the stability and regulations of the current grid.

A hybrid strategy, fusing grid-tied and off-grid systems components to produce flexible solutions, has recently gained popularity. Energy storage is included in hybrid systems to allow for the sale of excess energy back to the grid while also serving as a backup power source during grid outages. This strategy aims to provide the best of both worlds by balancing autonomy and connectivity.

In summary, there are several elements to consider when choosing between off-grid and grid-tied solar power systems, including geographic location, energy requirements, cost, and regulatory constraints. Every system has benefits and drawbacks, highlighting the significance of a customized strategy to satisfy specific needs. The interaction between off-grid and grid-tied systems highlights the dynamic landscape of renewable energy adoption, where flexibility and innovation pave the way for a cleaner, more resilient future as the community turns its attention to sustainable energy solutions.

Benefits and Challenges of Off-Grid Solar Power

Off-grid solar power systems have become revolutionary because they offer a way to achieve energy independence and supply electricity to places not serviced by conventional utility grids. Off-grid solar power's advantages and disadvantages reflect the dynamic interactions between technology, the environment, and socioeconomic variables shaping sustainable energy adoption.

The capacity of off-grid solar power to provide electricity to isolated and underserved places is one of its main advantages. Off-grid solar systems are a lifesaver in areas where connecting to the grid is either logistically or monetarily impractical. Solar energy can be used by isolated towns, off-grid homesteads, and remote cottages to meet their electricity demands, promoting economic growth and raising living standards. Off-grid solar power has a revolutionary effect beyond just providing electricity. It creates opportunities for communication, healthcare, and education, empowering communities in ways that go much beyond simply flipping a switch.

One of the main benefits of off-grid solar power systems is energy independence. Users become independent of centralized power infrastructure by producing their electricity. This independence is beneficial in places where the traditional grid's dependability is doubtful or when outages are expected. Off-grid systems with energy storage components, like batteries, guarantee a steady power supply without direct sunshine or outside disturbances. Off-grid solar power promotes resilience and self-sufficiency, which align with the larger objectives of sustainability and environmental care.

Off-grid solar energy also makes a substantial contribution to environmental preservation. In contrast to traditional energy sources dependent on fossil fuels, solar energy is clean, renewable, and produces no carbon emissions when it generates power. Off-grid technologies lessen the environmental impact of electricity consumption by lowering the carbon footprint of isolated areas. This alignment with sustainable practices is particularly relevant today when ecological degradation and climate change are major worldwide concerns. Off-grid solar power embodies the concepts of environmentally friendly energy solutions and provides a concrete and significant means of moving towards a low-carbon future.

A critical factor in the advantages of off-grid solar power is economics. Off-grid systems are now more economically feasible than ever because of government incentives and subsidies and the falling costs of solar technology. Long-term utility bill reductions cover the initial cost of solar panels and related components, particularly in areas where regular grid electricity is costly or unstable. Off-grid solar power systems provide an affordable substitute for supplying energy to isolated locations, alleviating the financial strain on people and communities that might otherwise find it difficult to obtain dependable electricity.

Nevertheless, there are obstacles to reaping the rewards of off-grid solar electricity. The initial installation cost is one of the main obstacles. Although the price of solar panels has decreased over time, off-grid solutions can require a sizable upfront expenditure for batteries and inverters. The problem is particularly acute in areas with low financial resources, as the viability of off-grid solar power depends on things like financing alternatives and government incentives.

The variable nature of sunshine presents a noteworthy obstacle for off-grid solar power installations. Off-grid systems rely on energy storage to fill the gap when solar generation is low, unlike grid-tied systems that can take power from the grid. This calls for careful planning and sizing of the system to guarantee that it can supply the energy needs even during protracted periods of overcast weather or low sunshine. Advancements in energy management systems and battery technology are addressing this difficulty. However, it still needs to be considered for the successful deployment of off-grid solar power in various locations.

Off-grid solar power users also face issues with maintenance and system management. Users must have a fundamental understanding of system maintenance because access to technical knowledge and spare parts may be restricted in distant places. Proactive maintenance, regular inspections, and prompt replacements are necessary for solar panels, batteries, and inverters to last a long time and function at their best. This feature emphasizes how crucial user support and education are to optimizing the advantages of off-grid solar energy.

Furthermore, in communities that are expanding or changing quickly, the scalability of off-grid solar power systems may provide a barrier. Increasing off-grid systems' capacity may necessitate investing in solar panels and storage as electricity consumption rises. A complex understanding of community growth and development is necessary to balance the system's capacity with the ever-changing energy needs.

Despite these obstacles, off-grid solar power is still becoming increasingly popular as a workable and sustainable solution. Technological advancements, in conjunction with policy incentives and community involvement, help surmount the challenges of implementing off-grid solar power. Initiatives and projects are growing to support the broader use of off-grid solar power as governments and organizations realize the socioeconomic and environmental benefits of illuminating distant locations.

In summary, the advantages and difficulties of off-grid solar power capture the intricate dynamics of adopting sustainable energy. The potential for off-grid solar power to revolutionize the energy industry is highlighted by its benefits for rural populations, energy independence, and environmental preservation. Although upkeep, intermittency, and upfront expenses are difficulties, further technological improvements and legislation that support them provide a path forward. Off-grid solar power illuminates the lives of people in even the most distant regions of the world, proving that access to electricity does not have to be limited to the grid. It can be obtained sustainably and equitably.

CHAPTER II

Assessing Your Energy Needs

Calculating Power Consumption

In contemporary times, where electronic devices and appliances play a significant role in our everyday lives, controlling and comprehending power consumption is essential to using energy responsibly and effectively. Learning electricity use is crucial for keeping tabs on utility costs and advancing environmental consciousness and sustainability. This section examines the complexities of power consumption calculations, reviewing basic ideas, measuring instruments, and real- world applications for people and companies.

Fundamentally, power consumption is the total quantity

of electrical energy a system or device uses in a certain amount of time. The watt (W) is the unit of measurement for Power, and to account for time, power consumption is commonly stated in kilowatt-hours (kWh). The calculation of power consumption can be done using the following simple formula: Voltage (V) x Current (A) equals Power (kW). This formula highlights the significance of voltage and current in determining a device's energy consumption by showing the direct link between voltage, current, and Power.

It is essential to comprehend how much Power each

item uses to manage energy consumption properly. Various instruments are available for this, from basic watt meters to more sophisticated energy monitoring systems. A simple yet useful tool, a watt meter plugs straight into an electrical outlet and shows the connected device's power consumption in real-time. Smart plugs with energy monitoring features enable users to remotely track and evaluate power consumption, providing insights into usage patterns and possible areas for optimization. This is a more thorough method.

Everyday domestic items like air conditioners, refrigerators, and light fixtures add a lot of Power to the total amount used. An appliance's rated power consumption, usually in watts on the item or in its documentation, must be considered when calculating its energy use. The energy consumption in kilowatt-hours is obtained by multiplying the rated Power (in kilowatts) by the number of hours the device operates. With the help of this straightforward computation, users may determine the energy consumption habits of certain appliances and spot areas where energy efficiency can be increased.

Energy monitoring systems are essential for measuring power consumption for enterprises, particularly those with complex and varied energy demands. These systems gather real-time energy consumption data across several departments or systems by integrating with the electrical infrastructure. The collected data can be examined to determine peak usage periods, evaluate the effectiveness of the machinery, and put load control plans into action. Accurately calculating and tracking power consumption is essential for satisfying sustainability targets and legal requirements in commercial and industrial environments with a significant energy footprint. It also helps keep costs under control.

Power consumption calculations require additional considerations for renewable energy systems, especially solar Power. Knowing how much Power a home or company uses is essential for constructing the right size solar installation in grid-tied solar systems, where excess energy can be recycled back into the grid. The capacity of the solar array required to supply most, if not all, of the energy requirements can be found by calculating power consumption. This process examines past energy bills, peak demand times, and prospective shifts in consumption patterns are evaluated.

The computations consider the power requirements of an entire building in addition to specific gadgets and appliances. Both residential and commercial buildings have a base load, the consistent, minimal amount of Power needed for necessities like lights, air conditioning, and electronic equipment in standby mode. Energy managers and homeowners can put waste reduction and energy optimization methods into practice by having a clear awareness of this baseline power consumption. This could entail investing in energy-saving appliances, upgrading insulation, or using intelligent technologies that provide more control over systems that use energy.

Even though power consumption calculations are an essential component of energy management, it's important to understand the broader ramifications and possibilities for improvement. Power consumption calculations yield information that enables people and organizations to make well-informed decisions that support energy efficiency objectives and foster a more sustainable future. Understanding energy use patterns encourages consumers to modify their behavior, such as turning off appliances when not in use, embracing energy-efficient technology, and putting load-shifting techniques into practice.

Furthermore, phantom loads—the ostensibly harmless power consumption of devices in standby mode—can be identified using an accurate power consumption calculator. Even when turned off, many electronic devices still need Power, which wastes energy. Users can target specific activities to reduce energy usage, including using smart power strips or unplugging gadgets while not in use, by quantifying the Power wasted by these phantom loads.

Calculating power consumption becomes crucial to corporate social responsibility and sustainability initiatives in industrial and commercial operations. Globally, businesses realize the value of environmental stewardship and implementing energy-saving strategies into their daily operations. Precise data on power use provides a starting point for establishing goals, monitoring advancement, and implementing plans to improve overall energy efficiency. Businesses that commit to lowering their energy footprint reap financial rewards and enhance their reputation by projecting an image of environmental consciousness to stakeholders and customers.

Calculating electricity consumption is difficult, even with the apparent benefits. A diverse strategy is necessary due to the complexity of devices and equipment, the dynamic nature of electricity demand, and the variability in energy usage patterns. The procedure is made more difficult by inconsistent data collection techniques, restricted access to real-time data, and a lack of defined metrics. To overcome these obstacles, industry players, legislators, and technology developers must work together to create shared guidelines, enhance data accessibility, and encourage cutting-edge monitoring tools.

The calculation and management of power usage could be revolutionized by integrating developing technologies like artificial intelligence (AI) and the Internet of Things (IoT). Smart devices with sensors and connections to IoT platforms make real-time data gathering and analysis possible. This provides a detailed understanding of patterns in energy usage. This enormous volume of data may be processed by AI algorithms, which can then spot patterns, abnormalities, and optimization opportunities. A paradigm for energy management that is more responsive and adaptable is made possible by the interaction of various technologies.

Finally, estimating power consumption is a key to sustainable energy usage, environmental stewardship, and responsible energy use. It goes beyond utility bills and numbers. The methods and instruments used to calculate power usage develop with technology, giving consumers ever-more-advanced insights into their energy-related behaviors. Power consumption measurement and analysis is a critical skill for creating a future in which energy sustainability and efficiency are not only ideals but essential elements of our everyday lives and shared responsibility, from individual homes to large industrial complexes.

Identifying Critical Loads

Understanding and ranking critical loads is essential for energy management since it guarantees electrical systems' dependable and effective operation. The vital parts that need a steady, uninterrupted power source to stay operational are known as critical loads. Identifying these loads is an essential first step when building robust power systems, especially in emergency response centers, data centers, and healthcare facilities where power outages can have serious repercussions. This ar investigates the notion of critical loads, looks at how to identify them, and considers the implications for making sure that essential services continue.

Critical loads are considered the electrical equipment and systems essential for maintaining basic operations in a particular setting. These loads might differ significantly according to the type of facility or activity. Life support systems, diagnostic tools, and refrigeration machines for medicine storage are examples of essential loads at a hospital. Critical loads in a data center include cooling systems to avoid overheating, networking equipment, and servers. Emergency response centers consider illumination, equipment, and communication systems critical loads. The realization that specific electrical loads are essential for upholding operational integrity and, in certain situations, sustaining life unites these disparate contexts.

Determining the critical loads in a system or facility requires carefully evaluating its relationships and operations. The first step in this process is to take a thorough inventory of every electrical load and classify it according to its significance for the business. A load's criticality is frequently assessed by considering its failure's effects on safety, health, the economy, and operational continuity. Critical loads are those that, in the event of an interruption, would have serious ramifications; non-critical loads, on the other hand, may withstand brief disruptions without jeopardizing the system's overall functionality.

One of the most essential tasks in healthcare settings is determining critical loads. The vital weights are medical imaging instruments, life support equipment, and diagnostic tools. Even a brief loss of power to these devices could have catastrophic effects on patient safety and health. Hospitals and other healthcare facilities install redundant power systems, backup generators, and uninterruptible power supply (UPS) units to provide a consistent and dependable power supply to vital loads. In the case of a grid breakdown, these techniques offer a smooth transition to other power sources, mitigating the dangers associated with power interruptions.

Similarly, it is crucial to identify essential loads in the context of data centers. The foundation of today's IT infrastructure comprises data centers, which house networking hardware, storage systems, and servers. These components must always be in operation for several internet services, organizations, and communication platforms to function. When a data center's critical loads are identified, innovative cooling solutions, backup generators, and redundant power systems can be installed to ensure ideal working conditions. The financial toll that data center outages exact highlights how important careful design and funding are to guarantee the resilience of vital loads.

Critical loads are vital to emergency response centers, which include police, fire, and disaster recovery units, to perform their essential tasks. Among the loads designated as critical are surveillance systems, emergency lighting, and communication systems. These facilities use backup power solutions like generators and UPS systems to provide an uninterrupted power supply and allow responders to coordinate, communicate, and carry out emergency operations even during a power outage.

Developing a thorough grasp of the power needs and dependencies of critical loads comes next after they have been identified. This evaluates each critical load's electrical properties, such as voltage, current, power factor, and operating time. It is also necessary to examine the interdependencies among essential loads to ensure the power system is built to support the simultaneous functioning of all the required equipment. With this degree of detail, backup power systems may be precisely sized and configured to satisfy critical loads' demands during brief and prolonged power outages.

Critical load identification becomes considerably more complex in the context of renewable energy, especially in off-grid or hybrid systems. When the power supply is erratic, batteries and other energy storage devices are used in these situations to close the gap between energy production and consumption. Prioritizing critical loads according to their significance, the power system's design ensures that their energy requirements are consistently met. Energy management systems are essential for tracking battery conditions, forecasting energy consumption, and distributing stored energy to critical loads as efficiently as possible.

The dependability and quality of the power supply also affect how resilient essential loads are. Sensitive electronic equipment can suffer lasting damage or malfunctions due to unstable voltage, frequency fluctuations, or power surges. Power conditioning devices like voltage regulators and surge protectors are frequently included in essential loads to provide a clean and steady power supply. By protecting vital loads from the damaging effects of electrical disruptions, these protective measures extend their lifespan and reliability.

Apart from technical aspects, determining critical loads necessitates a risk assessment considering exogenous circumstances like natural calamities, system malfunctions, or other catastrophes. By being aware of the weaknesses of essential loads in the event of an interruption, backup plans and redundant systems can be implemented to reduce downtime. Critical loads are guaranteed to continue operating even in unfavorable circumstances thanks to our comprehensive approach to risk management.

Beyond the immediate operational environment, successfully identifying and protecting essential loads has far-reaching repercussions. It results in better patient outcomes and increased safety in the healthcare system. It translates into uninterrupted service delivery and satisfied customers in data centers. It translates into prompt and efficient crisis interventions in emergency response centers. The wider socioeconomic ramifications highlight how crucial strategic planning and investment are to guarantee the resilience of vital loads.

Furthermore, the idea of critical loads is consistent with environmental stewardship and sustainability objectives. Reducing energy waste and optimizing power generating and storage systems for key activities are two benefits of effectively controlling and safeguarding critical loads. In addition to lowering costs, producing electricity lessens the environmental impact. Managing essential loads effectively is an important aspect of responsible energy use in a society where promoting sustainability and combating climate change are of utmost importance.

The resources available for determining and controlling essential loads change as technology progresses. Artificial intelligence and machine learning algorithms drive sophisticated monitoring and control systems that provide predictive maintenance for critical loads and real-time insights into power usage patterns. To reduce the possibility of unplanned breakdowns and maximize the effectiveness of crucial load operations, these technologies improve the proactive management of power systems.

To sum up, one of the most important aspects of energy management is recognizing and safeguarding critical loads. Whether in data centers, emergency response, healthcare, or renewable energy systems, the capacity to protect and prioritize critical electrical loads guarantees the uninterrupted operation of vital processes. The procedure entails deploying backup and redundant systems, taking a comprehensive approach to risk management, and thoroughly grasping the unique requirements of essential loads. Critical loads take center stage as society grows more dependent on energy for basic needs, highlighting the necessity of technical innovation, strategic planning, and a dedication to resilience in the face of uncertainty.

Seasonal Variations and Energy Storage

Variations in energy supply and demand brought about by the seasons present problems for power systems and emphasize the necessity for efficient energy storage technologies. Understanding and managing the effects of seasonal changes on energy generation and consumption become critical as the demand for renewable energy sources grows. The complex relationship between seasonal fluctuations and energy storage is examined in this section, along with its consequences for sustainability, grid stability, and the move toward a more dependable and resilient energy future.

Seasons significantly impact how much energy renewable energy sources like wind and solar power produce. Despite being plentiful and environmentally benign, these sources are cyclical by nature and weather-dependent. For example, the production of solar power peaks in the summer and decreases in the winter or areas with prolonged cloud cover. Similarly, wind power shows seasonal trends, with variations in wind speed impacting wind turbine output. The difficulty is in balancing this erratic energy output with the steady annual need for electricity.

Using energy storage becomes imperative in mitigating the disparity between fluctuating energy generation and consumption. When production is high, excess energy is generated; this energy can be stored in batteries, pumped hydro storage, or other storage technologies and released when production is low or demand is high. In areas with more noticeable seasonal fluctuations, including those with distinct summer and winter weather, the importance of energy storage is magnified. These technologies solve the issues brought on by the fluctuation of renewable energy supply, aid in integrating renewable energy sources into the current infrastructure, and support grid stability.

Battery storage is one well-known technology in energy storage, and it has made significant strides recently. Electric vehicles and grid-scale storage projects alike now turn on lithium-ion batteries, in particular, as the preferred method of storing and distributing electricity. Battery storage is a versatile and scalable solution that can effectively manage seasonal fluctuations in energy output. Batteries are used to store extra electricity generated during times when renewable energy production is at its highest. Then, when demand is high, or the output of renewable energy is low, this stored energy is released.

The electricity grid is made more resilient and flexible by adding battery storage. Batteries serve as a buffer, guaranteeing a steady power supply even when renewable energy generation is insufficient by storing extra energy during times of plenty. This is especially important in areas where seasonal variations in solar or wind resources affect the grid's overall stability. Systems for storing batteries offer a dynamic solution that adapts quickly to variations in energy consumption and strengthens the strength and resilience of the grid.

Pumped hydro storage is a proven and extensively used energy storage type beyond batteries. This technology uses the gravitational potential energy of water. Water is pumped to an elevated reservoir using surplus electricity generated during times of energy excess. The stored water is released when there is a great demand for energy, flowing downhill and turning turbines to create electricity. When overcoming the difficulties of seasonal changes, pumped hydro storage—which can store massive amounts of energy for extended periods—complements battery storage.

Energy storage can reduce the effects of seasonal changes in more ways than only electricity production. Energy storage solutions, like house batteries, allow users to optimize their energy use throughout the year when it comes to residential and business energy consumption. When solar production is low, or demand is higher, these systems release the extra energy produced by rooftop solar panels during the sunny months. This decentralized energy storage method improves total energy consumption efficiency, lessens dependency on the grid, and aids in load balancing.

Furthermore, the incorporation of seasonal energy storage impacts the more significant shift to a decarbonized energy system. Renewable energy is essential to the global effort to lessen the effects of climate change and reduce dependency on fossil fuels. To achieve a dependable and resilient energy infrastructure, seasonal changes must be addressed as renewable energy sources remain unpredictable. By guaranteeing the constant availability of clean energy, energy storage makes the integration of renewables easier and speeds up the shift to a sustainable energy future.

Although energy storage technologies present intriguing answers, there are obstacles to their general adoption. Large-scale energy storage initiatives, including constructing battery storage facilities or pumped hydro plants, can have significant up-front expenses. Through subsidies, tax breaks, and favorable regulatory frameworks, governments, legislators, and industry stakeholders are essential in encouraging and facilitating the adoption of energy storage systems. To fully utilize energy storage to mitigate seasonal changes and support the stability and sustainability of the electricity grid, these financial obstacles must be removed.

The geographical setting also influences the influence of seasonal fluctuations on energy generation and storage. Areas with more distinct seasons—those with severe winters or protracted cloud cover, for example—face particular difficulties. To provide a steady power supply in these places, efficient energy storage systems are even more essential. The complexity of the equation is further compounded by the seasonal variations in energy demand brought about by things like higher needs for heating or cooling. It becomes essential to customize energy storage systems to the unique features of each location to maximize their efficacy and guarantee grid stability throughout the year.

Research and development efforts are still being made to find novel approaches to seasonal energy storage as the need to address climate change and move toward sustainable energy systems grows. Innovative technologies that provide more alternatives for overcoming the difficulties caused by seasonal changes include flow batteries, compressed air energy storage, and thermal energy storage. These technologies offer several methods for storing and releasing energy, each with unique benefits and things to keep in mind. Seasonal variation management could become even more efficient, affordable, and scalable with the continued advancement of the energy storage environment.

In summary, converging energy storage and seasonal fluctuations signify a crucial turning point in pursuing a resilient and sustainable energy future. Addressing the erratic and unpredictable nature of renewable energy sources is critical as they become a more significant part of the world's energy mix. Energy storage technologies provide flexible ways to store excess energy during high-generation periods and release it during low-production or high-demand periods. Examples of these technologies are batteries and pumped hydro storage. The power grid's stability and dependability are guaranteed by the effective integration of seasonal energy storage, which also quickens the shift to a more decarbonized and sustainable energy ecology. The idea of a day when seasonal variations present our energy systems with few difficulties is getting closer to reality as long as research, innovation, and investment in energy storage continue to progress.

CHAPTER III

Designing Your Off-Grid Solar System

Choosing the Right Solar Panels

Solar energy stands out as a crucial component in the shift toward cleaner and more sustainable energy sources in the ever-changing renewable energy landscape. Selecting the appropriate solar panels becomes essential for people, companies, and governments wishing to capture the sun's power as the demand for solar energy rises. This section examines the many aspects to consider when choosing the best solar panels. It looks at the different kinds of solar technology, durability, efficiency parameters, and broader implications for the longevity and efficiency of solar power systems.

Photovoltaic (PV) panels, sometimes called solar panels, are gadgets that use the photovoltaic effect to turn sunlight into electricity. Numerous solar panel technologies are available on the market, each with unique qualities and uses. Based on the type of silicon used in their manufacturing, monocrystalline and polycrystalline solar panels are the two main varieties. Single-crystal silicon is used to create monocrystalline solar panels with a more homogeneous and effective structure. Compared to polycrystalline panels, these panels often offer better and more space-efficient efficiency rates. Polycrystalline solar panels, on the other hand, are more affordable but have a little lower efficiency since they employ several silicon crystals.

The decision between monocrystalline and polycrystalline solar panels is based on the particular needs of the project, the available funds, and the available space. Monocrystalline panels are frequently preferred for residential installations because they are more efficient and take up less space. Due to their affordability, polycrystalline panels are used in larger commercial installations where optimizing available space may be more critical than the panels' marginally poorer efficiency. It's important to remember that the differences between these panels are becoming increasingly hazy due to technology, and contemporary panels frequently provide competitive performance independent of crystalline structure.

Thin-film solar panels offer an alternative to crystalline technology. Photovoltaic material is deposited in thin layers onto a substrate, usually metal or glass, using thin-film technology. Because of their adaptability and flexibility, thin-film panels can be used when rigid crystalline panels are not feasible. Even though thin-film solar panels are often less efficient than crystalline ones, they work well in some situations, such as large-scale utility installations or solar-integrated building materials.

When selecting solar panels, efficiency is crucial since it directly affects how much electricity is produced for every unit of sunshine. More significant percentages of sunlight are usually converted into power by higher efficiency panels, which saves space and makes them appropriate for installations with limited ground or roof area. Higher efficiency, though, is frequently more expensive. Thus, it's critical to compromise budgetary constraints and efficiency. Determining the ideal efficiency for a particular solar power system involves evaluating the available space, the fiscal restrictions, and the specific energy needs.

When choosing solar panels, endurance and durability are essential considerations in addition to efficiency. Solar panels are subject to various environmental elements, such as sunshine, temperature changes, and meteorological conditions. Selecting panels with solid construction and materials resistant to these difficulties is essential. The frame, which is usually composed of aluminum, needs to be reliable and resistant to corrosion. Tempering the glass that covers the solar cells will increase its resilience and shield it from environmental stresses like wind and hail. Examining the manufacturer's warranty is another crucial way to determine how long and well the panel should last.

Their installation orientation and tilt influence solar panels' total efficiency and energy production. South-facing panels in the northern hemisphere receive the most fantastic sunshine during the day, which maximizes energy production. But the tilt angle is also critical. Optimizing the tilt angle according to the installation site's latitude enhances solar exposure. For instance, a steeper tilt angle at higher latitudes makes up for the reduced solar angle in the winter. To maximize the energy generation of solar panels, it is essential to understand the local climate, sunshine patterns, and how best to align and tilt the panels.

Aesthetic factors may also play a role in selecting solar panels for residential solar installations. Aesthetics are becoming a significant factor for homeowners as solar panels are increasingly incorporated into homes. In place of conventional rack-mounted panels, solar shingles and building-integrated photovoltaics (BIPV) offer design choices that fit in seamlessly with roofing materials. Even though these solutions cost a little more, they can provide a lot of value in aesthetics and integration with a home's overall design.

Choosing the appropriate solar panels also depends on the installation's location. Variations in the amount of sunlight received by different places throughout the year impact the solar panel's ability to produce energy. The solar potential of a given area can be estimated using tools like solar irradiance maps and online calculators, which consider factors like local weather patterns, shading, and sunshine hours. Throughout the system's lifespan, optimal performance and return on investment are ensured by selecting solar panels appropriate for the local sunlight conditions.

Local laws, government subsidies, and incentives influence the decision-making process when selecting solar panels. Many countries promote the use of solar electricity by providing tax breaks, rebates, or other financial incentives. Recognizing and taking advantage of these benefits can considerably reduce a solar power system's initial cost. Furthermore, adherence to local ordinances and building requirements is necessary to guarantee the lawful and secure installation of solar panels. Dealing with trustworthy solar installers knowledgeable about regional laws makes navigating the difficulties of compliance and permits easier.

Innovations that improve solar panel functioning and efficiency are being introduced by the rapidly developing field of solar technology. For example, bifacial solar panels use reflected light from surrounding objects to collect sunlight from the front and back of the panel. Energy output is increased by this technology, particularly in areas with reflective surfaces like snow, water, or light-colored rooftops. Single- or dual-axis tracking systems can optimize solar panel exposure by adjusting the panels' orientation to follow the sun's course. These technologies could be more expensive, but they could also increase energy yield; thus, they are worth looking into in some cases.

As the solar industry develops, factors other than producing electricity become more critical. A crucial component of the sustainability equation is the environmental impact of the manufacturing of solar panels, the recycling procedures, and the disposal of end-of-life materials. A more sustainable solar sector is facilitated by manufacturers who emphasize ecologically friendly procedures, employ recycled components, and offer choices for recycling or repurposing panels once their useful lives are over. Certifications like the Cradle demonstrate a dedication to environmentally responsible activities to Cradle accreditation or compliance with the Restriction of Hazardous Substances (RoHS) directive.

In summary, selecting the appropriate solar panels

necessitates carefully assessing several variables, each affecting a solar power system's effectiveness, performance, and long-term sustainability. Aesthetics, durability, efficiency rates, crystalline technology, location, and regulatory compliance are all factors considered throughout the selection process. Furthermore, other technologies that add new layers to the decision-making process include solar tracking systems, thin-film, and bifacial panels. Making wise and sustainable decisions along the path to solar power exploitation requires keeping up with technological developments, comprehending local solar conditions, and matching solar panel selections with personal needs and objectives. All of these things will become more and more important as the solar industry develops.

Battery Storage Solutions

Energy storage has gained attention because of the growing importance of renewable energy sources and the rising demand for decentralized power networks. Among the range of technologies, battery storage solutions have become crucial facilitators for grid stability, the effective integration of renewable energy, and the shift to a more resilient and sustainable energy landscape. This section explores the various applications, technology, and implications of battery storage and its impact on energy storage in the future.

The capacity to store electrical energy for later use is at the heart of battery storage systems. This feature helps mitigate the natural erratic nature of renewable energy sources such as wind and solar power by enabling extra energy storage during high-generation periods and releasing it during periods of low production or high demand. Batteries are essential in balancing out variations in energy supply and demand to maintain grid stability and reliability.

Different kinds of batteries meet other uses and needs. Lithium-ion (Li-ion) batteries have emerged as the standard technology for various uses, including electric cars, grid-scale energy storage, and portable gadgets. They are excellent choices for storing and distributing electricity due to their high energy density, efficiency, and comparatively extended cycle life. Lithium nickel manganese cobalt oxide (NMC) and lithium iron phosphate (LiFePO4) are two Li-ion battery variants that provide unique benefits and let consumers customize their selection according to budget, energy density, and safety needs.

Even though lead-acid batteries are an established and mature technology, they have uses in some situations. Lead-acid batteries are less expensive and more dependable than Li-ion batteries. However, they still have a lower energy density and a shorter cycle life, which makes them ideal for off-grid power solutions and uninterruptible power supply (UPS) systems. Advanced lead-acid technologies such as gel batteries and absorbent glass mats (AGMs) improve performance in some application circumstances.

Another type of energy storage technology that is gaining popularity is flow batteries, especially for grid and large-scale applications. These batteries' more modular and adaptable construction is made possible by the energy they store in liquid electrolytes. For instance, the benefit of decoupling power and energy capacity allows for flexibility in system design when using vanadium redox flow batteries. Flow batteries are being investigated for their possible contribution to improving grid resilience. They perform very well in situations requiring long-duration energy storage.

Battery storage systems are used in many industries, each with its own needs and advantages. Home energy storage systems allow homeowners to store extra energy produced by solar panels during the day to use at night or during times when solar production isn't as high. These devices increase energy independence, lessen dependency on the grid, and offer backup power in the event of a grid failure. The need for home charging solutions is fueled by the growing popularity of electric vehicles, which combine residential and transportation energy sectors to provide a positive outcome.

Commercial and industrial establishments use larger-scale battery storage to optimize energy use, lower peak demand costs, and improve energy efficiency. Businesses can better control their electricity expenses and meet sustainability and cost-saving objectives by installing behind-the-meter battery systems. Grid operators can also benefit from the assistance of grid-scale battery storage projects in reducing the impact of intermittent renewable energy sources, balancing supply and demand, and supplying ancillary services that improve grid stability.

One revolutionary use for battery storage is the idea of virtual power plants or VPPs. VPPs combine disparate energy resources, such as home solar and storage systems, as a centralized power source. This virtualized method maximizes the use of distributed energy assets, increases grid flexibility, and strengthens power supply reliability. VPPs demonstrate how battery storage solutions can completely change the traditional energy landscape by integrating decentralized resources into a cohesive and responsive network.

During catastrophes and natural disasters, the robustness and dependability provided by battery storage solutions become paramount. Energy storage-equipped microgrids give vital infrastructure a decentralized, independent power source, guaranteeing the continuation of essential services if the primary grid is disrupted. Battery storage improves community resilience and emergency response capabilities in areas vulnerable to catastrophic weather occurrences or grid failures.

One of the main factors influencing the adoption of battery storage options is their economic feasibility. Energy storage is now more widely available thanks to declining costs brought about by economies of scale, improvements in battery chemistry, and higher production efficiency. Government subsidies, incentives, and advantageous regulatory frameworks further encourage the implementation of battery storage projects. Research and development efforts are being made to improve batteries' performance, longevity, and sustainability as the industry develops. This will increase cost savings and expand the battery's application base.

Introducing battery storage into the energy system raises significant sustainability and environmental impact issues. However, using renewable energy is made possible by batteries, which significantly reduce greenhouse gas emissions, producing and disposing of batteries present ecological concerns. Reducing the environmental impact of energy storage systems requires sustainable battery production procedures, such as ethically sourced raw materials and recycling programs. A more sustainable battery lifecycle can be achieved by adopting circular economy ideas and developing closed-loop recycling solutions.

As we move toward a more sustainable and decentralized energy future, energy storage technologies are essential to the ongoing energy transition. However, issues and restrictions exist on how large battery storage solutions can be deployed. The problem of resource availability for crucial elements used in batteries, such as cobalt and lithium, is one significant challenge. To overcome these obstacles and guarantee the long-term sustainability of energy storage solutions, initiatives are being made to diversify the raw material supply chain, investigate alternate chemistries, and invest in recycling technology.

In summary, battery storage technologies are essential for a resilient and sustainable energy future. They are crucial for grid stability, energy independence, and disaster preparedness because of their adaptability, scalability, and capacity to handle the intermittent nature of renewable energy. Battery storage has many uses, from grid-scale projects to residential applications, highlighting how drastically they change the energy landscape. The energy ecosystem will become more efficient, dependable, and sustainable if battery storage technologies are developed and the sector adopts sustainability standards.

Inverters and Charge Controllers

An age of cleaner and more sustainable energy generation has begun with the impressive rise of renewable energy sources, especially solar and wind power. The essential parts of these systems are charge controllers and inverters. These gadgets are necessary for maximizing renewable energy systems' effectiveness, dependability, and performance. This section delves into the roles, varieties, and importance of charge controllers and inverters, elucidating the complex coordination required to facilitate the smooth assimilation of renewable energy sources into our electrical system.

Inverters provide a link between the alternating current (AC) power that powers most electrical grids and is utilized by our homes and businesses and renewable energy sources, which frequently produce direct current (DC) electricity. Their main job is transforming the DC electricity generated by wind turbines or solar panels into the AC electricity needed to run lights, appliances, and other electrical equipment.

Inverters are especially essential parts when it comes to solar energy. Photovoltaic (PV) solar panels produce DC electricity when exposed to sunlight. But AC is the type of energy that runs our homes and businesses. Thus, inverters make converting solar panels' DC output easier into AC power compatible with the grid and our everyday electrical needs.

Various kinds of inverters are intended for use in particular systems and applications. For example, string inverters are frequently utilized in household solar power systems. They are simple to use and reasonably priced, connecting several solar panels in series to transform their total DC output into AC power. However, each solar panel has a microinverter mounted, offering more precise control and may boost system efficiency.

Another technology that complements inverters is power optimizers, especially in setups where performance may be impacted by shade or different panel orientations. These gadgets maximize the output of every solar panel, guaranteeing that the system functions as efficiently as possible overall. Large volumes of DC power from numerous strings of solar panels are handled by central inverters, frequently employed in utility-scale solar projects. They transform the DC power into AC power for distribution to the grid.

Hybrid inverters are a recent development in inverter technology that enables integrating energy storage devices and converting DC to AC. When energy demand outpaces supply, these inverters will allow consumers to store excess energy produced during high renewable energy output. This skill is essential for achieving energy independence and resilience and providing a more dependable and consistent power supply.

Charge controllers are crucial protectors of energy storage, ensuring that batteries are charged and discharged correctly, while inverters have energy conversion from one form to another. Charge controllers function as clever gatekeepers in battery-storage renewable energy systems, controlling the flow of electricity to and from the batteries.

Preventing deep draining or overcharging batteries is a crucial role of charge controllers since it can negatively impact their longevity and performance. While severe draining can shorten the batteries' lifespan and capacity, overcharging can cause significant heat buildup and damage. Charge controllers keep an eye on the batteries charge level and modify the power flow to keep the batteries at the ideal amount of charging for a longer lifespan.

Maximum power point tracking (MPPT) and pulse width modulation (PWM) are the two primary categories of charge controllers. The more conventional and straightforward choice is PWM controls. Depending on how full the battery is, they adjust the charging current. Even though they work well, PWM controllers are less efficient than their MPPT counterparts, primarily when the solar panels operate at different sun intensity levels.

In contrast, MPPT charge controllers are more sophisticated and effective. They continuously modify the solar panels' electrical working point to maximize the amount of power available. The system's entire performance is improved due to this dynamic optimization, which guarantees that the solar panels function at their maximum efficiency under various environmental circumstances. Even though MPPT controllers are often more expensive, off-grid, and grid-tied, renewable energy systems prefer them because of their capacity to increase energy production.

Charge controllers are also essential when integrating energy storage with renewable energy sources. In off- grid solar projects, charge controllers govern battery charging and discharging to deliver a steady and dependable power source when the power grid is unavailable. Charge controllers and inverters work together in grid-tied battery-storage systems to maximize the use of stored energy, ensuring that it is used when grid electricity is expensive or unavailable.

The smooth functioning of renewable energy systems depends on the efficient coordination of charge controllers and inverters. Charge controllers play a minor role in grid-tied solar installations without energy storage because the energy produced is immediately consumed or sent back into the system. In these situations, inverters play a significant role in making sure that the electricity generated meets the demands of both consumers and the electrical grid.

However, the cooperation between charge controllers and inverters gets more complicated in off-grid and hybrid energy-storage systems. Batteries are used by off-grid systems, which are famous for standalone power solutions or in isolated locations, to store extra energy produced during sunny or windy days. When there is little or no renewable energy output, charge controllers optimize the batteries state of charge to fulfill energy demands.

Inverters and charge controllers must work harmoniously for hybrid systems, which integrate renewable energy sources, energy storage, and a grid link. Excess energy in these systems can be sold back to the grid, stored in batteries for later use, or used in emergencies when the grid is down. Inverters and charge controllers work together to manage energy flows, maximize self-consumption, and reduce dependency on outside sources.

The complementary effects of inverters and charge controllers become evident when considering distributed energy systems. The significance of these components becomes even more critical as the energy landscape moves towards decentralized models, where energy generation occurs at multiple places closer to consumers. For instance, homeowners who install domestic solar-plus-storage systems profit from the clever regulation of energy flows made possible by charge controllers and inverters. The extra energy produced during the day is saved for use at night, and the system runs smoothly to deliver a dependable and effective power source.

Furthermore, the capacities of renewable energy systems are improved by incorporating cutting-edge technologies like digital charge controllers and intelligent inverters. Ancillary services like voltage support and frequency regulation are provided by smart inverters, which help to stabilize the grid. Digital charge controllers with communication capabilities allow users to improve their energy consumption patterns and react to dynamic grid situations by enabling remote monitoring and management of energy storage systems.

The continuous development and integration of charge controllers and inverters are critical to the future of renewable energy systems. Research and development activities are concentrated on improving these components' scalability, intelligence, and efficiency as the market for energy storage and renewable energy solutions expands. New developments like bidirectional inverters, which facilitate smooth energy transfer between renewable energy sources, storage, and the grid, are opening the door to more advanced and durable energy systems.

CHAPTER IV

Installation Process

Site Assessment and Planning

The urgent need to reduce climate change and switch to sustainable energy sources has transformed the quest for renewable energy from a laudable goal to a global need. A thorough site assessment and planning process is essential to the success of any renewable energy project. This section explores the various factors that support the development of solar, wind, and other clean energy initiatives, emphasizing the crucial necessity of site evaluation and planning in renewable energy.

The first step in the site evaluation process for solar

energy projects is to have a complete grasp of the solar potential of the surrounding terrain. Solar irradiance, or the quantity of sunlight that reaches a given location, is a crucial factor that affects how practical and effective solar installations are. Quantifying solar radiation levels and determining the best sites for solar panels are made more straightforward with sophisticated tools like Geographic Information System (GIS) mapping and solar resource assessment software.

Site orientation and shading analysis are two essential

elements of evaluating a solar site. Azimuth, the term for the orientation of solar panels, is necessary for optimizing energy capture. Throughout the day, panels facing north in the southern hemisphere and south in the northern hemisphere receive the most sunshine. A shading analysis evaluates potential obstacles that could cast shadows on solar panels, such as structures, trees, or topographical factors. By reducing energy losses caused by shade, this study makes sure the solar array runs as efficiently as possible.

Furthermore, the tilt angle of solar panels affects energy generation, particularly in areas with significant seasonal changes. By adjusting the tilt angle according to the installation site's latitude, panels may maximize energy capture and capture as much sunlight as possible throughout the year. Developers can realize the full potential of solar energy resources by tailoring the orientation and tilt of solar panels to local conditions.

Site assessment for wind energy projects focuses on assessing the wind resource and comprehending the local wind regime. The performance and viability of wind turbines are mainly dependent on wind speed, direction, and fluctuation. Long-term anemometer readings yield essential information about the properties and variability of the wind resource.

An additional crucial component of evaluating wind sites is terrain analysis. The site's geography influences wind patterns. Therefore, developers must consider hills, valleys, and elevation variations. When choosing the best location and height for wind turbines, it's essential to take wind shear, how wind speed changes with altitude, and the degree of turbulence into account. Extra computational methods like Computational Fluid Dynamics (CFD) simulations may be needed to predict wind flow patterns on complex terrain effectively.

The process of micrositing, or choosing particular sites for individual wind turbines within a wind farm, entails a thorough analysis of the wind and terrain. To minimize wake effects—a condition in which downstream turbines experience decreased wind speeds due to the wake produced by upstream turbines—turbine spacing, layout, and orientation are optimized to maximize energy extraction. Micrositing prevents performance losses brought on by turbulence and wake effects, and the wind farm is guaranteed to run effectively.

Site evaluation goes beyond the features of solar and wind resources as energy storage integration becomes increasingly essential to renewable energy projects. The viability of integrating energy storage options, such as batteries, into the project is greatly influenced by variables including land availability, environmental concerns, and accessibility to the electrical grid.

Site conditions influence the decision between distributed and centralized energy storage systems. Centralized storage provides economies of scale but necessitates enough room and equipment; it is typically found at a single location within a renewable energy installation. Although distributed storage offers flexibility, it may have more significant implementation costs because it is scattered over the project site or incorporated with individual solar panels or wind turbines. These elements must be carefully considered during the site evaluation process to choose the best energy storage configuration.

Site planning also heavily relies on environmental factors, such as geology, soil conditions, and ecological impact evaluations. Sustainable procedures are essential for renewable energy projects' long-term sustainability and environmental stewardship. Examples of these practices include avoiding sensitive ecosystems and minimizing soil disturbance. The need to take a comprehensive approach to site assessment and planning is highlighted by the need to balance the advantages of clean energy generation with environmental conservation.

The regulatory environment must be navigated during site evaluation and planning, as it differs depending on the region and jurisdiction. For renewable energy projects to be successful, obtaining permissions and licenses is an essential step in the development process, and following local laws is crucial. Developers must interact with regulatory bodies, environmental agencies, and local communities to resolve complaints, obtain permits, and guarantee compliance with zoning and land-use restrictions.

Large-scale renewable energy projects typically require environmental impact studies or EIAs. These analyses examine possible effects on the environment, society, and culture, assisting stakeholders and decision-makers in making informed decisions. The permitting process necessitates public discussions, stakeholder participation, and transparent communication to resolve concerns about noise, visual impact, and other project-related factors and to build community support.

Furthermore, determining whether it is feasible to link renewable energy projects to the electrical grid requires conducting grid connection assessments. Grid capacity, infrastructure needs, and compatibility with current transmission or distribution networks influence the project's grid integration. Coordination with utility companies and grid operators is necessary to solve the technical, financial, and regulatory elements of grid connection and guarantee that the renewable energy project can successfully contribute electricity to the grid.

Optimizing the efficiency of renewable energy projects requires careful technology selection, system design, resource evaluation, and regulatory concerns. The selection of monocrystalline or polycrystalline solar panels affects the system's overall performance in solar projects. Energy capture and project economics are impacted by the hub height, rotor diameter, and wind turbine technology used for a project.

Hybrid renewable energy systems—which integrate several sources, such as solar and wind—need a sophisticated system design to manage energy supply and demand. Computer-aided design (CAD) and energy simulation software are examples of energy modeling tools that help evaluate how well hybrid systems function in different scenarios. Batteries and other energy storage technologies are included in the system architecture to improve flexibility, dependability, and the capacity to provide a steady power source.

Modern inverters and innovative grid technologies are examples of technological innovations that enhance system optimization by smoothly integrating renewable energy sources, increased grid stability, and grid support services. The ever-changing landscape of renewable energy technology emphasizes how critical it is to keep up with the latest developments and integrate them into site planning and design to optimize effectiveness and long-term profitability.

During the site evaluation and planning stages, precise financial modeling and risk analysis are critical to the economic sustainability of renewable energy projects. Developers must assess the project's return on investment, accounting for construction expenditures, ongoing operating costs, and prospective income sources such as sales of electricity, certificates of renewable energy, and subsidies or incentives.

The levelized cost of electricity (LCOE), a crucial indicator of the lifetime cost of electricity per unit, is evaluated by financial models. Sensitivity assessments assist in identifying potential risks and uncertainties by considering differences in important characteristics like energy production, equipment costs, and financing terms. Developers may attract investors, make well-informed judgments, and acquire funding for project implementation with the help of solid financial modeling.

A wide range of elements are included in risk assessments, such as market dynamics, resource variability, regulatory uncertainty, and technology hazards. Resilience and long-term project success depend on comprehending and reducing these risks. Developers are guided in navigating the uncertainties inherent in renewable energy production by a thorough risk assessment framework that includes contingency plans, risk management methods, and scenario studies.

Community participation is an essential component of the site evaluation and planning process to promote cooperation and increase local support for renewable energy projects. Effective communication with stakeholders, indigenous groups, and local communities is crucial to resolve issues, get input, and ensure the project is in line with community interests and values.

Community benefits, including job creation, regional economic growth, and educational programs, can be incorporated into the project's design to maximize the project's beneficial effects on the neighborhood. They are establishing open lines of communication, holding public gatherings, and making information easily accessible, all of which help to promote trust and a feeling of community among locals.

To sum up, planning and site assessment are essential components of a successful renewable energy project. Developers need to thoroughly evaluate the natural resources, regulatory environment, and community dynamics before using solar, wind, or a combination of both. The trajectory of renewable energy programs is shaped by the complex interplay among environmental considerations, technology choices, regulatory compliance, and financial feasibility. Developers may set the path for resilient, community-supported, and sustainable renewable energy projects that contribute to a cleaner and more sustainable energy future by adopting a holistic approach that considers various elements and involves stakeholders at every stage.

Installing Solar Panels

Putting in solar panels is a concrete and significant step toward using clean, renewable energy. Solar energy becomes more important as the world struggles to move away from fossil fuels and lessen the effects of climate change. This is because a sustainable future is becoming more and more critical. This section delves into the complexities of solar panel installation, covering everything from the planning and site preparation stages to the technical requirements of mounting the panels and connecting them to the grid. We can learn more about how communities, companies, and individuals can actively contribute to a more resilient and sustainable energy landscape by demystifying the installation process.

Setting realistic goals and doing a comprehensive feasibility evaluation are necessary before installing solar panels. This entails assessing the site's solar potential, comprehending the solar irradiation levels in the area, and considering elements like space availability, orientation, and shadowing. Sophisticated instruments, such as online calculators and maps of solar resources, can determine how much electricity a solar system can produce given certain geographic and meteorological conditions.

An essential first step in the planning process is establishing specific goals. Whether the goal is to become energy independent, save power costs, or improve environmental sustainability, figuring out these goals helps with solar panel system design and sizing. Whereas commercial or industrial projects emphasize large-scale energy generation and cost savings, residential installations might concentrate on offsetting a portion of domestic electricity consumption.

Preparing a site so solar panels can efficiently capture sunlight entails creating the right environment. This can entail removing vegetation, leveling the ground, and ensuring adequate drainage for ground-mounted solar arrays. To ensure that a building can sustain the added weight of solar panels, it is imperative to evaluate the structural integrity of rooftops.

The solar panels' tilt and orientation are important considerations to maximize energy extraction. Panels facing south receive the most incredible sunlight in the Northern Hemisphere, and maximum exposure is ensured by adjusting the tilt angle according to latitude. Shading analysis reduces energy losses from shadows by considering obstructions such as surrounding structures or trees.

Selecting the best solar panels requires careful

consideration of aspects, including efficiency, technology, and design. Common possibilities include monocrystalline and polycrystalline panels, which are differentiated by the type of silicon used in their production. Because of their single-crystal structure and better efficiency rates, monocrystalline panels are a good choice for installations with limited space. Although slightly less efficient than monocrystalline panels, polycrystalline panels are still quite affordable. They are made up of many silicon crystals.

An additional option is thin-film solar panels, which

deposit small layers of photovoltaic material onto a substrate. Thin-film panels provide flexibility and suppleness, which makes them acceptable for specific applications where rigid panels might not be viable, even if they are often less efficient than crystalline choices. The visual appearance of solar installations is becoming more and more critical. Solar shingles and building-integrated photovoltaics (BIPV) offer aesthetically pleasing substitutes for conventional rack-mounted panels.

Inverters act as a link between the alternating current

(AC) power utilized in homes and businesses and the direct current (DC) electricity produced by solar panels. The system architecture and the particular project must determine which inverters to use—string, micro, or power optimizers. Multiple solar panels are connected in series with string inverters to convert their aggregate DC output to AC power. In contrast, microinverters are put on a panel-by-panel basis, providing more precise control and possibly increasing system efficiency.

Power optimizers maximize each panel's output and

reduce shadowing impacts by collaborating with string inverters. Installing inverters requires careful electrical work, attention to safety regulations, and consideration of elements like temperature management to ensure optimal performance. With developments like hybrid inverters that make it easier to integrate energy storage systems and allow households to store extra energy for later use, inverter technology is still advancing.

A crucial part of the installation process, physical solar panel mounting and racking ensures stability, longevity, and maximum sun exposure. A sturdy racking system firmly fastens the panels to the earth when installing panels on the ground. The panels' orientation and angle are meticulously adjusted to match the best possible solar exposure, frequently with tracking systems that trace the sun's path throughout the day.

Mounting systems on rooftops must consider the load-bearing capability and weatherproofing specifications of the structure. Water leaks can only be avoided by adequately flashing and sealing the areas surrounding roof penetrations. Flat roofs frequently use ballasted systems, which offer stability due to the weight of the panels and racking. Installers fix the panels on pitched roofs without compromising the roof's structural integrity by using brackets or rails.

The solar power system's components must be assembled during the electrical wiring and connection phase to route the electricity produced safely. Alternating current (AC) wiring runs from the inverter to the main electrical panel, while direct current (DC) wiring links the solar panels to the inverter. Installers ensure the system complies with electrical rules and standards by adhering to safety procedures, such as using conduits and properly grounding the system.

For solar installations that are related to the electrical grid, connecting to it is an essential first step. These systems enable net metering and possible financial incentives by allowing surplus electricity produced by the solar panels to be sent back into the grid. Coordination with the neighborhood utility and adherence to interconnection guidelines are necessary for grid connection. A charge controller and energy storage system —usually batteries—are included in the wiring configuration in off-grid installations, where solar panels serve as the primary power source to offer a steady and dependable power supply.

The last phase of installing solar panels is commissioning, during which the system is put through a thorough testing and verification procedure to ensure it is operating correctly. This entails performing a rigorous system assessment, confirming the functionality of inverters, and inspecting electrical connections. In grid-tied installations, this stage involves finalizing the connection to the electric grid.

Monitoring systems, which are frequently coupled with inverters, let businesses, households, and installers keep tabs on how well solar power systems work. Up-to-date information on energy output, system performance, and possible problems facilitates prompt repair and improves system management. By helping to optimize patterns of energy consumption, monitoring also enables users to make well-informed choices regarding energy use and possible efficiency gains.

Maintaining solar installations over the long term requires proactive maintenance and a dedication to environmental responsibility. Even though solar panels require very little upkeep, maximum performance requires routine cleaning, inspections, and sometimes repairs. Monitoring systems are essential for spotting irregularities or drops in energy output so that prompt actions can be taken.

Environmental factors also apply to solar panels nearing the end of their useful lives. Even though solar panels last at least 25 years, recycling programs are becoming increasingly popular to dispose of defunct panels. Recyclable materials can be recovered with minimal environmental impact because of technological developments in recycling, and conscientious businesses provide recycling programs.

To sum up, installing solar panels is a life-changing step toward environmentally conscious living and sustainable energy. Every stage of the process, from the preliminary evaluations of viability and goal-setting to the last phases of commissioning and monitoring, calls for meticulous thought, knowledge, and a dedication to quality. Installing solar panels is a concrete and significant step toward a cleaner, more sustainable, and resilient future as solar technology develops and the world understands the need for renewable energy.

Connecting Batteries and Inverters

When it comes to finding sustainable energy solutions, the combination of batteries and inverters is a powerful combination that facilitates the efficient use of renewable energy sources. The necessity for dependable energy storage has grown as the globe adopts more sustainable energy sources. The symbiotic relationship between inverters, which convert the direct current (DC) into the alternating current (AC) used in homes and businesses, and batteries, which can store surplus energy, improves energy storage systems' efficiency, resilience, and adaptability. The complexities of connecting batteries and inverters are examined in this section, which also provides insights into the uses, prospects, and technical aspects of energy storage.

The wide range of battery technologies is at the heart of the revolution in energy storage. The performance and features of energy storage systems are significantly impacted by the type of battery used, which ranges from conventional lead-acid batteries to cutting-edge lithium-ion and new technologies like flow batteries. Lithium-ion batteries, especially, have become industry leaders in several applications because of their high energy density, efficiency, and extended cycle life.

The first step in connecting batteries and inverters is to comprehend the unique specifications and features of the selected battery technology. The main factors that affect how batteries and inverters interact are voltage, capacity, and discharge characteristics. Batteries in a networked system hold onto extra energy produced during peak production times by renewable energy sources like solar or wind power. When production is low, or demand is high, this stored energy provides a valuable resource that improves the overall stability and dependability of the energy system.

Inverters are essential components of the energy storage ecosystem because they make it easier for DC-powered batteries to be seamlessly integrated with AC-based electrical grids and appliances. Inverters' primary job is to transform DC electricity stored in batteries into AC power, bringing it into compliance with traditional electricity distribution regulations. The particular application, the size of the system, and the necessary functionality all influence the choice of inverter technology.

In-home solar systems, string inverters, microinverters, and power optimizers are frequently used to control the DC-to-AC conversion of solar panels and establish a grid connection. Hybrid inverters become crucial when batteries are included in the mix. In addition to converting DC to AC, hybrid inverters also manage battery charging and discharging, maximizing energy flows and raising the energy storage system's overall efficiency.

Grid-forming inverters and other advanced inverter technologies have become more common in microgrids and grid-tied systems. These inverters can regulate voltage and frequency independently, providing support and stability to the grid. When batteries are added, their job becomes even more crucial since they control the two-way flow of electricity between energy storage systems, renewable energy sources, and the grid.

Compatibility must be carefully considered, especially voltage and capacity when connecting batteries and inverters. Inverters must comply with these specifications to guarantee effective and secure operation. Batteries are intended to function within particular voltage ranges. Furthermore, the battery's capacity—expressed in megawatt-hours (MWh) or kilowatt-hours (kWh)—determines how much energy it can hold and release.

A standard lithium-ion battery in a home might have a voltage of 48 volts, but more extensive systems, like those in commercial or industrial settings, might have higher voltages. The voltage of the attached batteries must be matched when choosing or configuring inverters. Compatibility with capacity is also crucial since, to prevent damage or performance deterioration, an inverter needs to manage the rates of discharge and charge that the battery system demands.

Manufacturers frequently offer inverters made especially for use with particular battery models or technologies, allowing easy integration. This compatibility streamlines the installation process and lowers the possibility of technical problems while guaranteeing peak performance and functionality. Advanced inverters with adaptive controls and intelligent management systems become crucial for balancing the various properties of the connected batteries when varied battery chemistries or capacities are involved.

Batteries and inverters work together to orchestrate a sophisticated ballet that governs the dynamics of charging and discharging. Inverters control DC power flow into the batteries during charging, whether it comes from the grid or renewable energy sources. During this procedure, the incoming DC electricity is transformed into the proper voltage and current levels needed to charge the batteries. The conversion or round-trip efficiency is an essential component in evaluating the energy storage system's overall performance.

Discharging, however, entails releasing the energy stored in batteries to power electrical loads or replenish the system with electricity. An inverter is an essential component in transforming battery energy from DC to AC power that satisfies grid voltage and frequency requirements. With its smooth integration into the current electrical infrastructure, this AC power can support local loads and help maintain grid stability.

Sophisticated energy management systems enable complex control over the charging and discharging procedures; these systems are frequently built into hybrid inverters. These systems optimize energy flows by considering variables, including battery conditions, grid demand, and electricity costs. Inverters can prioritize self-consumption in grid-tied systems by storing extra energy produced during low demand and discharging it at high demand, resulting in cost savings and supporting the grid.

Connecting batteries and inverters allows for peak shaving and load shifting, two significant advantages. Peak shaving is the practice of utilizing stored energy to lessen reliance on the grid during times of peak electrical demand. In this case, when electricity costs rise, batteries drain, saving users money and reducing peak demand, which helps maintain system stability.

The idea is expanded upon by load shifting, which deliberately controls the timing of energy storage and release. Energy consumers can maximize their electricity savings by charging batteries during off-peak hours, when electricity rates are lower, and discharging them during peak demand. This dynamic control aligns with the larger objective of demand-side management in the electrical grid and is made possible by advanced inverter and energy management systems.

Peak shaving and load shifting are two factors that help energy storage systems remain economically viable and appealing for use in commercial and industrial settings. Apart from reducing expenses, these tactics improve the adaptability of energy consumers by offering a dependable energy supply amid power disruptions or elevated electricity rates. Connected batteries and inverters provide flexibility that turns energy storage systems into valuable assets that can adjust to the ever-changing electricity markets.

Beyond localized advantages, battery and inverter integration provides grid support and auxiliary services. The ancillary services that competent inverter-equipped grid-tied energy storage systems can offer to improve the grid's stability and dependability. Reactive power compensation, voltage control, and frequency regulation are all included in this support.

Grid-forming inverters are essential for improving the resilience of energy infrastructure since they can function independently and provide stable grid conditions. These inverters can be cut off from the grid during blackouts, yet they can still power essential loads and sustain small-scale microgrids locally. This capacity improves the overall dependability of the electrical supply, particularly in areas that heavily rely on renewable energy sources or are prone to disturbances.

CHAPTER V

Maintenance and Troubleshooting

Regular Maintenance Practices

Regular maintenance techniques are becoming increasingly important as the world moves toward a renewable energy future. The effective operation of renewable energy systems, such as solar panels that collect sunlight, wind turbines that harness wind power, and energy storage systems that store excess electricity, depends on proactive and regular maintenance. This section explores the importance of routine maintenance procedures in maintaining the longevity and efficiency of renewable energy systems. It also looks at the advantages of proactive maintenance, the changing landscape of maintenance procedures in the renewable energy industry, and essential factors to consider when evaluating different technologies.

Relying on photovoltaic (PV) technology, solar energy systems comprise a significant share of the renewable energy market. Maintaining solar panels regularly is essential to their longevity and best performance. One of the critical maintenance chores is cleaning the solar panels to remove accumulated dust, filth, and debris that can obstruct their ability to absorb sunlight. This is especially important in arid areas or places that see seasonal variations, as these climatic conditions might affect the effectiveness of solar panels.

Inspections for potential damage, such as cracks or scratches on the solar panels, are fundamental to identifying issues early on. If damage is ignored, it could jeopardize the system's overall performance. In addition, routine inspections are necessary for electrical parts like wiring and inverters to guarantee correct operation and avert any safety risks. Frequent output monitoring of the system, frequently made possible by sophisticated monitoring software, makes it possible to identify performance deviations from expectations and take appropriate action quickly.

Wind energy captured by tall wind turbines also requires careful maintenance procedures to maintain effectiveness. The constant forces of wind and weather can cause wear and tear, erosion, or damage to the turbine blades over time. Therefore, routine inspections are essential. Modern imaging drones are becoming proper instruments for comprehensive airborne inspections, giving a complete picture of the turbine's state without requiring a lot of downtime.

Apart from visual examinations, periodic checks are necessary for mechanical parts like gearboxes and bearings to identify wear indicators and guarantee proper lubrication. Essential parts of wind turbine maintenance include tightening bolts, checking electrical connections for integrity, and evaluating safety system operation. Addressing concerns proactively boosts the turbine's efficiency and contributes to safety by decreasing the chance of component failures.

Energy storage systems are essential for counteracting the intermittent nature of renewable energy sources. These systems frequently include cutting-edge battery technologies. Maintaining batteries properly is crucial to navigating their complex chemistry and extending their lives. Frequent observation of temperature, voltage, and charge state aids in spotting unusual activity or any problems with the battery system. With real-time information from advanced energy management systems, operators may decide on charging and discharging cycles with knowledge.

Because excessive temperatures can influence battery performance and safety, battery enclosures must have adequate ventilation and temperature management. Periodic inspections of cooling systems, if applicable, and thermal control mechanisms enhance the energy storage system's overall dependability. Furthermore, regular calibration and capacity tests support reliable state-of-health evaluations, enabling operators to detect and correct any decline in battery performance.

The growing prevalence of hybrid renewable energy systems that include solar, wind, and energy storage means that maintenance procedures must change to accommodate interactions between various parts. Maintaining solar panels, wind turbines, batteries, and inverters harmoniously requires a comprehensive maintenance strategy. Inspections of the entire system, including mechanical and electrical parts, are necessary to find any possible dependencies or interactions affecting overall performance.

Predictive maintenance procedures are made more accessible and smoother by integrating modern control systems and communication protocols. For instance, data analytics and machine learning algorithms can analyze previous performance data to identify possible problems and recommend preventive actions. Predictive maintenance reduces downtime while improving the hybrid renewable energy systems' overall dependability and economics.

Proactive maintenance techniques have several advantages for operators and owners of renewable energy systems. Proactive maintenance enhances returns on investment by maximizing system efficiency and reducing downtime, which is its most important feature. Early problem identification and resolution lowers the need for expensive repairs or component replacements by preventing minor difficulties from growing into more significant failures.

Furthermore, preventive maintenance helps to ensure that renewable energy systems are dependable and safe. Finding and fixing such risks, like electrical malfunctions or structural problems, protects the environment around the system as well. Proactive maintenance helps improve grid stability and resilience in grid-tied systems, which are prominent in integrating renewable energy with the electrical grid.

Proactive maintenance is intrinsically linked to the lifespan of renewable energy installations. Periodic inspections and maintenance prolong the life of various parts, such as batteries, inverters, solar panels, and wind turbine blades. This lessens the environmental effect of production and disposal, delaying the need for significant capital expenditures while supporting the industry's larger sustainability objectives.

Even with the clear advantages of proactive maintenance, there are still difficulties in understanding the maintenance paradigm for renewable energy systems. The lack of industry-wide standard maintenance procedures and the variety of technology present some problems. Every technology, including energy storage, wind, and solar, has different factors to consider and needs a different maintenance strategy.

Maintenance teams need help due to the remote locations of several renewable energy projects. Planning is necessary, and specialized equipment is frequently required to reach off-grid energy storage systems in remote locations, solar arrays spread over vast deserts, and wind turbines positioned atop hills. New technologies can alleviate these issues and reduce physical presence requirements, such as robotic equipment for automated inspections or remote monitoring systems.

An increasing focus is being placed on integrating digital technologies to transform maintenance processes as the renewable energy industry grows. Intelligent maintenance strategies are created by data analytics, artificial intelligence, and the Internet of Things (IoT). Operators are empowered to make well-informed decisions and manage resources efficiently through remote diagnostics, predictive analytics, and real-time monitoring.

Additionally, developments in sensor technology offer insightful information about the state of components. Early anomaly identification is made possible by continuously monitoring factors, such as temperature fluctuations in solar panels or wind turbine vibration levels. With the move toward condition-based monitoring, maintenance tasks can now be planned according to the actual condition of the machinery, maximizing resources and reducing needless interventions.

Predictive maintenance is a growing concept that uses machine learning algorithms to predict possible faults based on inputs from the real world and previous data. By using a proactive rather than reactive approach to maintenance, operators can solve problems before they influence system performance. When it comes to vital parts, like the gearboxes in wind turbines or the battery cells in energy storage systems, predictive maintenance is essential since timely repair can avert expensive malfunctions.

Routine maintenance procedures are the cornerstone of effective and sustainable renewable energy systems. It is impossible to overestimate the importance of maintenance in maintaining these systems' robustness, security, and lifespan as the globe depends more and more on sustainable energy sources. From the massive wind turbines scattered across the countryside to the photovoltaic fields collecting solar energy and from the complex energy storage systems regulating supply and

Realizing the full potential of renewable energy systems requires proactive maintenance, which is defined by routine inspections, preventive interventions, and the incorporation of cutting-edge technologies. While there are still difficulties, digital transformation, predictive analytics, and a dedication to sustainability are characteristics of the changing landscape of maintenance operations.

Promoting cooperation amongst industry players, standardizing maintenance procedures, and embracing cutting-edge technologies will be crucial as the renewable energy sector develops. In addition to serving as evidence of prudent resource management, creating a sustainable maintenance framework is essential in ensuring future generations have access to cleaner, more resilient energy.

Monitoring System Performance

Monitoring system performance becomes crucial for sustainable operations in the dynamic world of renewable energy, where solar arrays spread over landscapes, wind turbines tower over horizons, and energy storage systems hum with stored electricity. Ensuring the optimal functioning of particular components and gaining insights into broader trends are benefits of effective monitoring, which help system operators and owners make well-informed decisions. This section explores the importance of tracking system performance for renewable energy sources—these critical metrics and how monitoring systems are changing to achieve sustainability need to be considered.

Photovoltaic (PV) panels, which use solar energy, depend on the continuous conversion of sunlight into electrical power. Monitoring system performance in solar systems becomes essential to maximize energy yield, spot possible problems, and improve overall efficiency. Solar irradiance, a crucial characteristic being examined, gauges the amount of sunshine that reaches the solar panels. Energy output is directly impacted by changes in irradiance brought on by clouds covering or shadowing.

Monitoring systems keep tabs on the efficiency of individual solar panels and inverters in addition to sun radiation. Environmental conditions can cause photovoltaic cells to deteriorate over time, and abnormalities like cracks or hotspots can reduce the cells' overall efficiency. Because they transform direct current (DC) into proper alternating current (AC), inverters are essential parts that require ongoing attention. Keeping an eye on their operation gives you information about voltage levels, efficiency, and other problems that could affect the solar energy system.

Real-time data capture is a common feature of modern solar installation monitoring systems, enabling operators to evaluate performance from a distance. Predictive maintenance and energy output can be optimized with the integration of machine learning algorithms and advanced analytics. As the solar business develops further, data-driven insights gained from extensive monitoring become essential for solar energy operations that are both sustainable and efficient.

Monitoring system performance is essential for navigating the efficient winds of wind energy, where turbines use the kinetic energy of the wind to create electricity. The first set of metrics covered by wind turbine monitoring is wind direction and speed. These factors directly impact the turbine's power output, and real-time monitoring ensures that the turbines are running within the wind speed range intended for maximum efficiency and safety.

The health of turbine blades is another focus point of monitoring. Aerodynamic efficiency and overall turbine performance can be compromised by wear and tear that results in erosion or damage. Advanced monitoring systems use sensors and imaging technology, such as drones outfitted with high-resolution cameras, to conduct exhaustive inspections. These checks help with predictive maintenance plans and identify apparent problems, enabling operators to handle possible blade problems before they influence performance.

The mechanical parts of wind turbines, such as gearboxes and bearings, are also monitored. Temperature checks, vibration analysis, and oil condition monitoring are essential for determining how well these parts are doing. Continuous monitoring enables early anomaly detection, resulting in proactive maintenance that minimizes downtime and extends the life of vital turbine components.

Energy storage systems add another complexity to the monitoring environment since they frequently use cutting-edge battery technologies. Tracking characteristics like the batteries' state of charge (SOC), state of health (SOH), and state of function (SOF) is part of monitoring the energy storage system's performance. These measurements shed light on the energy storage system's present capacity, general condition, and operational capabilities.

Monitoring the charging and discharging cycles in real-time is essential to maximize the efficiency of energy storage systems. Monitoring systems keep tabs on how well batteries charge and discharge, guaranteeing that energy is transferred with the most minor loss possible. This process is further improved by integrating energy management systems and intelligent inverters, enabling energy flows' dynamic control and coordination.

An essential component of battery performance and safety is temperature control. Monitoring systems measure the temperature of battery cells and enclosures continually, allowing for the activation of heating or cooling devices as needed. Temperature anomalies can indicate possible problems, and operators can take corrective action based on real-time monitoring, extending the life and dependability of the energy storage system.

Monitoring in hybrid renewable energy systems—which integrate solar, wind, and energy storage—becomes a symphony of interdependent parts that must be harmonized for best results. The interplay among solar panels, wind turbines, batteries, and inverters demands that system performance be monitored holistically. Smooth coordination is made possible by sophisticated control systems and communication protocols, which give operators insight into the interdependencies and synergies of various components.

Data integration and analytics become crucial in hybrid systems because real-time decision-making relies on fully understanding the interconnections between multiple renewable energy sources. Monitoring systems provide predictive maintenance, load forecasting, and adaptive control strategies by offering a unified picture of the complete system. The importance of monitoring in maintaining the smooth integration of disparate components will only increase with the growing prevalence of hybrid systems.

Understanding the difficulties involved in this attempt and maintaining a sharp focus on essential criteria are necessary for navigating the monitoring environment of renewable energy systems. Solar energy systems must monitor solar irradiance, panel efficiency, and inverter performance. Wind energy requires continuous attention to wind speed, turbine blade health, and mechanical component quality. Temperature, charging and discharging cycles, and battery parameters must be monitored in energy storage devices. The difficulty with hybrid systems is combining and analyzing data from various sources to get insightful conclusions.

The geographical dispersion of installations, the variety of renewable energy technologies, and the requirement for uniform monitoring techniques present challenges for system performance monitoring. Interoperability problems between monitoring systems and components made by different manufacturers might result from the need for standard protocols. Furthermore, many renewable energy facilities are located in remote areas, which makes on-site maintenance and inspections logistically tricky.

Some of these problems are addressed by emerging technologies like the Internet of Things (IoT). Remote monitoring and control are made possible by real-time connectivity and data transfer provided by IoT-enabled sensors and devices. A more cohesive monitoring ecosystem is made possible by standardization initiatives in communication protocols, data formats, and cybersecurity measures, which facilitate the seamless integration of various systems and components.

Predictive analytics and digital technology are bringing about a significant change in the monitoring system performance landscape. The digitization of renewable energy operations lessens the need for manual interventions by enabling real-time monitoring, data analytics, and remote control. A new age in intelligent renewable energy systems is being ushered in by the Industrial Internet of Things (IIoT) with the deployment of sensors, actuators, and communication devices.

Machine learning algorithms enable predictive analytics, improving monitoring systems' performance. These algorithms foresee possible problems, spot trends, and recommend preventive actions based on past performance data analysis. By addressing any issues before they arise, predictive maintenance techniques based on data-driven insights help operators minimize downtime and maximize the overall efficiency of renewable energy systems.

Cloud-based monitoring tools make centralized access to data from dispersed renewable energy installations possible. This makes managing many systems easier and allows for more extensive use of data-driven decision-making. Operators can contribute to the continuous advancement of renewable energy technology by honing operating tactics, enhancing system designs, and utilizing knowledge from many installations.

Finally, it becomes clear that monitoring system

performance is essential to guaranteeing the robust and long-lasting functioning of renewable energy systems. The capacity to collect real-time data, evaluate performance parameters, and make deft decisions is crucial for every renewable energy system, including solar arrays, wind turbines, energy storage systems, and hybrid configurations. Predictive analytics and the continuous digital revolution are changing the monitoring scene and providing a window into a future where renewable energy systems function with previously unheard-of efficiency and dependability.

Prioritizing standardized monitoring procedures,

encouraging industry stakeholder participation, and funding research and development will be crucial as the renewable energy sector grows. In addition to serving as evidence of ethical resource management, the transition to resilient and sustainable operations is essential in ensuring future generations have access to cleaner and more sustainable energy sources.

Common Issues and Troubleshooting Tips

Renewable energy systems, including solar, wind, and energy storage technologies, are critical to the global transition toward sustainable energy sources. They are not impervious to difficulties and problems that could compromise their functionality; nevertheless, they are much like any other complicated technical system. It is essential to comprehend prevalent issues and possess efficacious troubleshooting techniques to guarantee the dependable functioning of renewable energy infrastructure. In addition to examining some common problems with solar, wind, and energy storage systems, this post offers advice on how to solve the issues and reduce their impact.

Regarding solar energy, photovoltaic (PV) panels and the performance of related components are common problems. Reduced energy output is a common problem caused by several reasons, like dirt collection, shadowing, or the gradual degradation of photovoltaic cells. To solve this problem, every solar panel must be carefully inspected, any shading components must be found and addressed, the panels must be cleaned regularly, and wear and tear must be periodically checked.

Another possible cause of issues is inverters, which transform the DC electricity produced by solar panels into useful AC power. The solar energy system may completely shut down due to inverter problems. Inverter troubleshooting includes monitoring the devices' efficiency, looking for unusually high heat levels, and resolving any output irregularities related to voltage or current. Maintaining the dependability of the inverter components can be aided by routine firmware upgrades and replacing older units' inverters.

Grid-tied solar systems have a problem from intermittent electricity output, sometimes called the "cloud effect." This happens when sun irradiance is momentarily reduced by passing clouds, causing variations in energy output. Advanced inverter technologies and energy storage solutions can be used to lessen this problem by lowering the unpredictability of energy production and giving the grid a more steady supply of electricity.

Wind energy systems have unique difficulties since they rely on the wind's kinetic energy to produce electricity. Mechanical problems, especially in the gearbox and bearings, are common in wind turbines and result in downtime and decreased energy output. To identify early indicators of wear or approaching failures, regular inspections, vibration analysis, and oil condition monitoring are necessary for troubleshooting these issues. Proactive maintenance is required to increase the longevity of vital turbine components, such as lubrication changes and component replacements.

Another frequent issue with wind turbines is blade damage, which is frequently brought about by external conditions, lightning strikes, or manufacturing flaws. It takes in-depth inspections with drones, visual examinations, or even sophisticated imaging technologies to find and fix blade problems. Early discovery enables prompt replacements or repairs, limiting additional damage and maximizing the turbine's aerodynamic performance.

Wind energy has an inherent problem in the form of variable wind speed, which affects power output. Variations in wind direction and turbulence can cause variations in energy production. These problems can be lessened with sophisticated control systems and yaw mechanisms, which change the turbine's orientation to face the wind. Operators can also predict changes in wind patterns and adjust turbine operations accordingly by integrating wind forecasting systems.

The intermittent nature of renewable energy sources makes energy storage devices necessary but also presents integration and battery performance issues. Degradation of capacity, decreased efficiency, and concerns with thermal control are standard. Monitoring battery metrics like state of charge (SOC) and state of health (SOH), doing routine capacity checks, and implementing efficient heat management techniques are all part of troubleshooting.

Over time, cyclic fatigue—a consequence of repeated cycles of charging and discharging—may cause a decrease in battery performance. Battery lifespan can be increased by implementing tactics like depth of discharge optimization and intelligent charging and discharging algorithms. Moreover, cyclic fatigue problems are addressed by battery technology improvements, such as more robust materials and better chemistries.

Problems could occur when the inverter in a grid-tied energy storage system synchronizes with the electrical grid. Frequency variations, voltage fluctuations, or grid disturbances may impact the stability of the energy storage system. Ensuring the inverter has grid-forming capabilities—which enable it to function independently and support grid stability during disruptions—is a crucial step in troubleshooting. Advanced inverters with grid-support features are essential to meet these problems.

Integrating various renewable energy sources, such as solar, wind, and energy storage, in hybrid systems presents issues with the smooth coordination of disparate components. Typical problems include inadequate power sharing across various sources, which results in ineffective use of available energy. To ensure that all system components operate balanced and integrated, troubleshooting these problems requires optimizing coordination and control methods.

In hybrid systems, poor communication between various subsystems can impair system performance. A thorough grasp of the data exchange and communication protocols between solar inverters, wind turbines, and energy storage components is necessary for troubleshooting. Integration and troubleshooting procedures can go more smoothly using standardized communication interfaces and protocols.

Because hybrid systems sometimes consist of AC and DC components, problems relating to voltage or frequency mismatches may occur. Ensuring inverters and converters are configured correctly to meet the overall system needs is a crucial step in troubleshooting. The difficulties associated with voltage and frequency coordination can be overcome by applying sophisticated power electronics and synchronization technologies.

A comprehensive methodology that considers both system-wide obstacles and specific component problems is needed to troubleshoot renewable energy systems effectively. Predictive analytics, sensor-based monitoring, visual inspections, and other routine maintenance procedures are vital in spotting possible issues before they become more serious. Preventive maintenance, like lubricant modifications, firmware upgrades, and component substitutions, adds to renewable energy systems' overall dependability and durability.

Operators can troubleshoot with precision thanks to diagnostic technologies, including data analytics for performance trend analysis, vibration analysis for mechanical faults, and infrared thermography for hotspot identification. Implementing a comprehensive and meticulously documented maintenance plan guarantees a systematic and thorough troubleshooting process that covers all essential components of the renewable energy system.

Effective troubleshooting requires operators and maintenance staff to receive education and training. A proactive troubleshooting technique involves knowing each technology's nuances, conversing with system architecture and integration, and keeping up with industry best practices. Maintenance staff can be better equipped to handle unforeseen obstacles with the help of training programs that offer practical experience and realistic scenarios.

Successful troubleshooting also requires cooperation with technology providers and equipment makers. Building a solid relationship enables operators to use the manufacturer's experience in solving particular problems, get timely information on equipment performance, and get technical support. Troubleshooting procedures and instructions suggested by the manufacturer provide essential insights into effectively addressing issues.

It is critical to keep lines of communication open with utility providers in the context of grid-tied systems. Participating in grid support programs, adhering to regulatory standards, and comprehending grid codes all help to ensure that renewable energy systems are seamlessly integrated with the more extensive electrical infrastructure. Working with utility providers can improve grid stability by making troubleshooting during disruptions easier.

Resolving typical problems with renewable energy

systems is essential to maintaining their dependable and long-term functioning. From solar arrays and wind turbines to energy storage systems and hybrid configurations, each technology brings distinct issues that necessitate a systematic and proactive approach.

To troubleshoot. Building resilience and reliability in

renewable energy systems involves implementing regular maintenance procedures, using diagnostic technologies, funding training initiatives, and encouraging industry partnerships.

The development of troubleshooting solutions will be

influenced by data analytics, technological breakthroughs, and the combined knowledge of industry professionals as the renewable energy sector grows. The dedication to troubleshooting excellence speeds up the shift to a cleaner and more sustainable energy future while protecting investments in renewable energy infrastructure.

CHAPTER VI

Maximizing Energy Efficiency

Energy Conservation Strategies

Energy conservation has become a key component of global sustainability efforts due to rising energy demand and the need to combat climate change. The prudent and efficient use of energy resources is known as energy conservation, and it includes a variety of tactics meant to cut down on energy use, minimize waste, and promote the switch to greener and more sustainable energy sources. This section examines the complex field of energy conservation methods, looking at approaches from different industries, the impact of innovation and technology, and the more significant ramifications for economic resilience and environmental stewardship.

Energy-saving techniques are essential for fostering

sustainable behaviors and lowering carbon footprints in homes. A crucial first step in reducing household energy use is the incorporation of energy-efficient appliances, such as LED lighting, smart thermostats, and energy-star-rated devices. By identifying inefficient areas, home energy audits help homeowners fix insulation gaps, stop leaks, and maximize the effectiveness of their heating, ventilation, and air conditioning (HVAC) systems.

Modifications in behavior are just as crucial for

preserving energy in homes. Over time, little habits like using natural sunshine, shutting off lights and appliances when unused, and heating water with awareness can result in substantial energy savings. By empowering people to make knowledgeable decisions about their energy use, educational efforts that promote energy literacy help communities develop a conservation mindset.

Adopting renewable energy in residential settings is a game-changing tactic in the fight against energy waste. By installing energy storage devices and rooftop solar panels, households can produce and store their electricity, decreasing their dependency on conventional grid-based power sources. Subsidies and incentive schemes encourage the use of renewable technologies even more, fostering a mutually beneficial relationship between specific conservation initiatives and more general sustainability objectives.

Energy conservation methods in the industrial sector are centered on eliminating energy-intensive operations and streamlining manufacturing processes. Efficiency evaluations and energy audits are fundamental instruments used by enterprises to pinpoint areas in need of development. Adopting sophisticated control systems, making process changes, and upgrading to energy-efficient machinery all help save significant energy without sacrificing output.

Cogeneration, or combined heat and power (CHP) systems, are an example of an integrated industrial energy conservation strategy. CHP systems improve overall efficiency and lessen dependency on separate power and heating systems by concurrently producing energy and using waste heat for heating or cooling processes. Industrial energy conservation efforts are further refined through energy management systems and real-time monitoring technologies, which offer practical insights for ongoing development.

Green building design is a paradigm shift in the construction industry where energy conservation is integrated into the structure itself. Energy-efficient HVAC systems, passive design ideas, and sustainable building materials all help lower the energy needed for lighting, heating, and cooling. A sustainable approach to urban development is fostered by certifications such as

LEED (Leadership in Energy and Environmental Design), which direct architects and builders toward environmentally sensitive practices.

Energy conservation initiatives are centered on the transportation sector, which accounts for many of the world's energy consumption and emissions. There has been a significant shift toward greener mobility with the rise of electric cars, hybrid technologies, and fuel-efficient vehicles. Government subsidies and developments in battery technology are driving up the uptake of electric vehicles (EVs) and decreasing reliance on internal combustion engines.

Energy conservation goes beyond electrification and includes public transportation marketing and intelligent urban planning to optimize transportation systems. Urban environments that consume less energy rely less on personal vehicles and integrate mass transit systems and those with infrastructure for cycling and pedestrians.

Fuel efficiency innovations are not limited to the automotive sector; they are also found in the aviation and maritime sectors. The aviation field is changing due to advancements in aircraft design, the use of sustainable aviation fuels, and the investigation of electric or hybrid propulsion systems. Similarly, improvements in ship design and the investigation of alternative fuels are assisting in reducing energy consumption in maritime transportation.

Innovations in technology are essential for promoting energy saving in various industries. The emergence of the Internet of Things (IoT) has facilitated the creation of energy management systems and smart grids that maximize energy distribution, minimize transmission losses, and provide consumers with real-time data on energy consumption. Automation technology and intelligent sensors improve industrial process efficiency by providing accurate control and reducing waste.

The development of smart home systems in building technology enables homeowners to monitor and manage energy consumption remotely. Dynamic energy saving is facilitated by smart thermostats, lighting systems, and linked appliances that modify their settings according to user preferences, weather, and occupancy. These systems are further refined by incorporating artificial intelligence (AI), which makes predictive analytics and adaptive energy management possible.

In addition to providing clean energy, renewable energy technology like wind turbines and solar photovoltaics also encourage energy conservation. These technologies are becoming increasingly competitive as alternatives to traditional fossil fuels because of their constant advancement and cost reduction. Advanced battery technology and other energy storage options make integrating intermittent renewable energy sources into the grid easier, resulting in stability and a decrease in the demand for backup power from conventional, less efficient sources.

The development and execution of efficient policy and regulatory frameworks play a pivotal role in igniting energy-saving endeavors. Throughout the world, governments are passing laws that set emission reduction objectives, encourage the use of renewable energy sources, and provide incentives for energy efficiency improvements. A legislative environment that promotes energy conservation includes carbon pricing systems, building regulations that require energy-efficient designs, and energy performance standards for cars and appliances.

Financial incentives, including grants, tax credits, and subsidies, encourage energy-saving activities in various industries. A culture of energy-conscious decision-making is fostered by incentive programs for household solar installations, energy-efficient retrofits in commercial buildings, and research and development projects in clean energy technology. These programs offer real benefits to businesses and individuals alike.

International cooperation and accords, like the Paris Agreement, highlight the world's commitment to halting climate change by reducing energy use and switching to renewable energy sources. Nations cooperate to tackle the common problem of lowering global energy consumption and lessening the effects of climate change by establishing shared goals, exchanging best practices, and encouraging technological transfer.

Initiatives to conserve energy have substantial financial ramifications in addition to environmental care. Energy security is improved, jobs are created, and innovation is encouraged when funds are allocated to energy efficiency and renewable energy initiatives. The shift to a low-carbon economy creates new opportunities for clean technology and services, promoting economic diversification and entrepreneurship.

Building retrofitting, installing smart grid infrastructure, and growing renewable energy projects all help create jobs for people with different skill levels. Specifically, the renewable energy industry's manufacturing, installation, maintenance, and research and development sectors have grown to be significant employers. Retraining and upskilling programs are crucial as the energy environment changes because they give workers the skills they need for a sustainable future.

Energy conservation also directly impacts energy affordability, particularly for disadvantaged groups. By lowering energy costs and increasing the accessibility of clean energy, efficient technology, and conservation practices serve to lessen energy poverty. Mainly, energy-efficient homes help to reduce utility bills, which benefits low-income households and advances social justice.

Smart Appliances and Devices

With the advent of the Internet of Things (IoT) and swift technical progress, intelligent appliances and gadgets have become indispensable components of modern life. These clever, networked gadgets are made to maximize energy efficiency, improve convenience, and give consumers unmatched control over their living environments. This section examines the complex world of smart appliances and gadgets, going over their development, how artificial intelligence (AI) has been incorporated into them, how this has affected energy efficiency, and how this technological revolution has affected society.

The journey towards smart appliances started with the incorporation of basic programmable features that let users personalize their devices' settings or set timers. But the real revolution came when connection enabled appliances to talk to consumers and one other over the internet. The foundation of smart homes is their connectivity, which allows products to work together harmoniously to produce automated and integrated living spaces.

Regarding kitchen equipment, smart refrigerators are a prime example of how conventional gadgets have developed into intelligent machines. These refrigerators, equipped with sensors, cameras, and touchscreens, can offer real-time inventory monitoring, recipe suggestions based on goods on hand, and even the ability for users to observe what's inside their refrigerator remotely. The user experience is further improved by incorporating voice recognition and AI-powered personal assistants, transforming the kitchen into a center of connected efficiency.

Similarly, by learning user preferences, modifying settings depending on occupancy patterns, and optimizing energy usage, smart thermostats have entirely changed how people regulate the temperature in their homes. Beyond controlling the temperature, these gadgets also provide energy usage information and personalized energy-saving suggestions and support the creation of more comfortable and sustainable living spaces.

Incorporating artificial intelligence has revolutionized the smart appliance industry by enabling it to learn from, adjust to, and anticipate customer requirements. Machine learning algorithms allow gadgets to customize their features to fit certain lifestyles by analyzing user behavior, preferences, and usage patterns. This degree of customization optimizes resource use based on real-time data, which improves customer pleasure while also helping to reduce energy consumption.

For example, AI is used by smart home security systems to differentiate between routine domestic tasks and possible security risks. Facial recognition technology allows cameras to detect authorized users and family members, only raising an alarm when unknown or suspect activity occurs. This adaptive filtering improves security by reducing false alarms and needless energy consumption associated with continuous monitoring.

Artificial intelligence (AI) algorithms allow intelligent lighting systems to adjust their lighting in response to occupancy, natural light levels, and user preferences. Through the dynamic adjustment of brightness and color temperature, these systems provide a visually pleasant and comfortable environment and help save energy. Bright lighting improves home security and energy economy by learning and imitating user behavior, such as repeating lighting patterns when on vacation.

The ability of intelligent appliances and gadgets to promote sustainability and energy efficiency is one of its main advantages. By reacting to real-time data, modifying settings in response to usage trends, and even taking part in demand response programs, connected devices can optimize energy consumption. With this degree of dynamic control, consumers may reduce their environmental footprint without compromising ease of use.

For instance, occupancy sensors, meteorological predictions, and learning algorithms are used by smart thermostats to provide customized heating and cooling plans. These gadgets improve user comfort, help lower energy costs, and minimize carbon emissions by not using extra energy when a space is empty. An additional layer of energy efficiency is added by remotely adjusting thermostat settings, enabling users to optimize settings before leaving for home.

Regarding smart home lighting, energy-efficient LED lights and clever controllers can save much energy. Automated lighting systems can create personalized lighting scenarios for various activities, adjust brightness based on natural light levels, and switch off lights in empty areas. By integrating motion sensors, lights are only turned on when necessary, reducing energy waste and promoting sustainability objectives.

Despite the allure of smart appliances, there are obstacles to their widespread adoption. Privacy and safety issues are brought up by security concerns, which include the possibility of unwanted access to linked devices or data breaches. To fix vulnerabilities and safeguard user data, manufacturers and developers need to prioritize strong cybersecurity measures, put encryption procedures into place, and update software often.

Another area for improvement is interoperability, which arises when customers integrate gadgets from various manufacturers into their networks of smart homes. Standardization initiatives, such as creating common communication protocols and compatibility standards, are crucial to guarantee smooth integration and communication across various intelligent appliances. A lack of compatibility may impede the entire potential of an automated and networked smart home.

User acceptance and awareness heavily influence the success of intelligent appliances. While tech-savvy people might welcome these improvements immediately, education and proof of the real benefits—such as energy savings, increased convenience, and improved quality of life—are needed for widespread acceptance. The initial financial barrier must be removed, and issues with complexity and data privacy must be resolved if smart appliances are to become widely accepted.

The widespread use of smart appliances, which are changing lifestyles, fosters a new degree of connectedness between people and their living environments. Users' Expectations and behavior have changed due to how convenient it is to monitor and manage home appliances from a distance using voice commands or smartphones. Once limited to science fiction, the idea of the "smart home" is now a real thing that affects how people interact with and view their homes.

Additionally, intelligent appliances are essential for advancing sustainability and environmental consciousness. These gadgets enable users to make environmentally conscious decisions by providing real-time feedback on energy consumption, proposing energy-saving solutions, and aiding in overall resource management. It is possible to track and evaluate energy usage patterns using the data produced by smart appliances, which will help both individuals and policymakers make well-informed decisions.

The introduction of intelligent appliances fits well with more significant cultural movements supporting sustainable and green living. Smart gadgets give people the tools to actively participate in energy conservation efforts as they become more aware of their environmental impact. With their ability to reduce wasteful energy use and facilitate the integration of renewable energy sources, intelligent appliances help us all move closer to a more sustainable and environmentally friendly future.

The future of smart appliances and gadgets will be one of constant innovation and development. Even more advanced and context-aware gadgets will be made possible by the confluence of technologies like edge computing, IoT, and artificial intelligence. More sophisticated sensors, better connectivity, and quicker computing power will all help to make user interfaces better and use less energy.

Integrating smart appliances into energy management systems and developing smart grids might produce more dynamic and responsive energy networks. Demand-side management, in which appliances can intelligently modify their operation in response to pricing signals and grid conditions, becomes a vital component of a more sustainable and robust energy infrastructure.

The influence of smart appliances is further amplified by the growth of innovative city programs and the development of networked ecosystems. Smart devices and urban infrastructure will change how cities function and use resources, from intelligent waste management systems to vehicle-to-vehicle communication systems. The idea of an innovative and sustainable city aims to combine efficiency, technology, and environmental protection harmoniously.

Integrating Off-Grid Systems with Other Renewable Sources

Off-grid systems have advanced significantly due to the need for sustainable energy and independence. These systems provide decentralized solutions that utilize renewable energy sources. Even though off-grid systems —frequently connected to freestanding power setups— have grown in popularity, combining them with other renewable energy sources is becoming more widely acknowledged. This section thoroughly analyzes how combining off-grid systems with solar, wind, and other renewable energy sources can improve energy efficiency, environmental effect, and reliability. It also discusses the advantages and disadvantages of doing so.

Off-grid systems are becoming a mainstay of energy solutions in isolated or difficult-to-reach areas because of their capacity to function independently of the centralized power grid. These systems often rely on solar panels, wind turbines, or a combination of renewable sources to create electricity, usually stored in batteries for later use. Off-grid solutions are appealing because they promote energy independence and resilience by supplying electricity to places with little or no access to traditional grid infrastructure.

Off-grid systems frequently use solar photovoltaic (PV) panels, which use sunshine to create electricity. These solar panels convert sunlight into direct current (DC), stored in batteries, or transformed into alternating current (AC) for instant usage. The panels are composed of linked solar cells. In contrast, wind turbines use a generator to convert the wind's kinetic energy into electrical energy. Both technologies are productive in providing clean and sustainable power solutions in stand-alone off-grid applications.

Solar energy is a natural ally for off-grid systems because it is abundant and widely available. Integrating solar energy and off-grid solutions improves the sustainability and dependability of energy generation. Solar panels are a common primary energy source in off-grid systems because they offer a reliable and steady power source. The synergistic relationship between solar power and off-grid systems is especially noticeable in areas with high solar irradiance, where sunlight is abundant for a considerable amount of the year.

Combining solar energy with off-grid systems has several benefits; one is the capacity to store energy produced during the day for use when there is little or no sunlight. Batteries and other energy storage technologies are essential for maintaining a balance between the supply and demand of energy. The efficiency and dependability of energy storage in off-grid applications have increased because of advancements in battery technologies, such as lithium-ion and flow batteries, guaranteeing a steady power supply even when renewable sources are not producing electricity.

Furthermore, the modular design of solar PV systems permits scalability in off-grid configurations. Off-grid systems can readily incorporate more solar panels to accommodate changing needs as energy demand rises or new power sources become available. This scalability is especially helpful in off-grid applications where increasing the number of homes with access to energy is a continuous objective, like rural electrification efforts.

When combined with off-grid systems, wind energy, defined by its unpredictability and reliance on local wind conditions, offers both potential and challenges. Because they are made to harness the wind's kinetic energy, wind turbines provide an additional energy source that can improve the overall dependability of off-grid systems. Combining the best features of solar and wind technologies, hybrid systems overcome the intermittent nature of renewable energy generation and offer a more reliable power source.

Diversifying energy sources is a significant benefit of combining wind energy with off-grid technologies. Even if there is a lot of solar energy available during the day, wind energy can also be obtained at night by increasing wind speeds. By reducing the influence of weather- related changes, this diversity ensures a more reliable and steady energy supply. Solar and wind technologies work together to provide a reliable and well-rounded renewable energy source for off-grid homes in windy regions.

Wind energy drawbacks, such as turbine noise and

visual impact, can be lessened in off-grid applications where residential neighborhood proximity isn't as problematic. Furthermore, vertical-axis and small-scale wind turbine technology improvements increase wind energy's viability for off-grid deployments. These turbines are ideal for decentralized applications because of their compact and modular design, facilitating seamless integration with pre-existing off-grid systems.

Hybrid systems are generally developed by combining

numerous renewable sources with off-grid systems. Hybrid systems optimize energy generation and improve overall system performance by combining various renewable energy technologies, such as solar, wind, and occasionally small-scale hydropower. These systems provide a consistent and dependable power supply by utilizing the advantages of each energy source.

Regarding renewable energy, hybrid off-grid systems

are a prime example of the symbiotic concept. Compared to solo solar or wind installations, these systems can offer a more reliable and constant power supply by intelligently controlling the generation and storage of electricity from many sources. Using sophisticated algorithms, energy management systems optimize efficiency and dependability by coordinating the operation of solar panels, wind turbines, and energy storage components.

Hybrid off-grid systems are especially helpful in isolated or island communities since they solve the problems associated with energy access and lessen reliance on imported fossil fuels. These systems provide a specialized and sustainable solution by being able to be adjusted to the site's unique energy requirements and resources. Because of their adaptability to changing conditions, hybrid systems are appropriate for various locations and applications.

Off-grid system integration with other renewable energy sources has great potential, but there are also drawbacks. The variable nature of renewable energy sources—mainly solar and wind—is one of the main factors to consider. Since these sources are sporadic, reliable energy storage must be found to guarantee a steady power flow. Developments in battery technologies, pumped-storage hydropower, and other storage techniques are essential to overcome this obstacle.

Additional hurdles in integration come from interconnection requirements and compatibility issues between various renewable energy technologies. For integrated off-grid systems to function as best they can, solar panels, wind turbines, and energy storage components must be able to interact with one another and work together harmoniously. Standardization initiatives within the renewable energy sector are essential to solving these issues and promoting interoperability.

Another factor to consider is the strong and durable infrastructure requirement in off-grid applications, especially in harsh or isolated locations. The performance of integrated systems depends critically on the dependability of the energy storage, control, and power electronics components. Off-grid installations must be regularly monitored to detect and resolve possible problems and guarantee long-term survival.

Integrating off-grid systems with other sources will be critical to the shift to decentralized energy solutions as technology continues to transform the renewable energy landscape. Integrating digitalization and automation, innovative grid development, and the continuous advancement of energy storage technologies are anticipated to augment the dependability and efficiency of integrated off-grid systems.

Communities and individuals can take charge of their energy production and use with the help of decentralized energy solutions. Because of the flexibility that integrated off-grid systems provide, customized solutions that can be adjusted to the particular requirements and features of various locales are possible. Integrated off-grid systems offer a flexible and scalable approach to localized energy generation for multiple locations, from isolated communities seeking access to energy to environmentally conscious resorts striving for sustainability.

CHAPTER VII

Off-Grid Living and Sustainability

Off-Grid Lifestyle Considerations

Self-sufficiency, sustainability, and a stronger connection to nature have drawn some to an off-grid existence. Living off-grid is appealing, but it has distinct obstacles. This section examines the complexities of off- grid living, from energy and water management to psychological and social issues.

Off-grid living is about energy independence, frequently achieved through renewable energy. Solar panels, wind turbines, and hydropower dominate electricity generation. These sources provide sustainable alternatives but require careful planning.

Off-grid electricity systems rely on photovoltaic panels. Solar panel efficiency and reliability depend on location, sunshine exposure, and weather. Off-grinders must evaluate their energy needs and construct a lifestyle-compatible solar power system. Energy storage, usually batteries, is needed to bridge the gap between energy output and consumption, especially during low sunshine.

Wind turbines use wind energy to generate energy. Wind power systems work best in places with consistent, moderate wind speeds. Wind energy improves durability in off-grid settings, compensating for solar power variability. Wind turbines require strategic location, regular maintenance, and off-grid living noise considerations.

Hydropower from flowing water is another off-grid energy option. Access to streams or rivers allows small- scale hydroelectric systems to generate electricity consistently. Hydropower requires knowledge of local hydrology and sustainability, like other renewable sources.

Energy conservation and efficiency are key off-grid.

Energy-efficient appliances, intelligent energy management, and conscious consumption become daily habits. Modern conveniences and off-grid electricity resources must be balanced strategically.

In the off-grid lifestyle, water becomes a vital resource.

Off-grinders use wells, rainwater, or adjacent waterways instead of municipal water. Sustainable water management is a daily concern.

Environmentally friendly rainwater collecting systems

use roof-mounted collection surfaces and storage tanks. When it rains, off-grid aficionados store rainwater for dry months. Filtration and purification make captured rainwater appropriate for home uses.

Off-grid properties get water from aquifers via wells.

However, well water quality must be monitored routinely, and extraction rates must be regulated to avoid depletion. Off-grinders employ renewable energy- powered pumping equipment to obtain water for home and irrigation usage.

Hydropower systems can deliver water to areas with

streams or rivers. Dual-use systems generate energy and ensure water supply for off-grid residences near flowing water sources. Compliance with legislation and environmental considerations ensures sustainable water resource management.

Off-grid living requires wastewater management. Composting toilets, greywater systems, and green sanitation reduce environmental impact. Off-grinders live in harmony with nature by using water responsibly and recycling and treating it.

Off-grid living is unique and rewarding, but it has its drawbacks. Becoming self-sufficient requires a mentality shift and a willingness to face uncertainty. Off-grid living is dynamic due to unpredictable weather, seasonal energy production, and proactive maintenance.

Off-grinders struggle with seasonal changes, especially in harsh areas. Winter temperatures and less sunlight can impair solar and wind energy generation. Planning for seasonal fluctuations requires improving energy storage, preserving resources, and having contingency plans to manage temporary shortages.

Off-grid living needs manual upkeep and troubleshooting. Off-grinders can quickly resolve challenges by understanding renewable energy, water supply, and trash management. Maintenance, inspections, and proactive system updates extend the life of off-grid infrastructure.

Off-grid living requires thought. Some find solace in a distant, off-grid property, escaping city life. However, solitude and restricted social interactions may be issues. Off-grinders must combine isolation and community interaction according to their principles and inclinations.

Sustainable, energy-efficient, and environmentally conscious building approaches are used to develop off-grid homes. In eco-friendly off-grid construction, passive solar design regulates temperature and lighting using natural factors.

Layout and orientation are critical to passive solar architecture. Structures that capture maximum sunlight in winter and shade in summer help natural heating and cooling. Large windows are carefully placed to maximize daylight and eliminate artificial lighting, saving energy.

Thermal mass, essential to passive solar architecture, requires heat-retaining materials. These materials store heat throughout the day and release it slowly at night. Common materials like adobe rammed earth, and concrete creates a more stable and comfortable indoor climate without active heating or cooling.

Insulation, a crucial component of energy-efficient design, helps off-grid residences retain heat throughout colder seasons and remain calm in hotter months. Natural insulating materials like sheep's wool, repurposed denim, and straw bales fit off-grid lifestyles and reduce construction's environmental impact.

Off-grid construction is more sustainable with renewable materials and local resources. Sustainable wood, recycled steel, and recovered building materials create eco-friendly structures. These materials promote a circular economy and reduce environmental effects.

This lifestyle generally includes food production, self-sufficiency, energy, and water. Organic gardening, permaculture, and small-scale agriculture are essential to off-grid living.

Off-grinders can grow fresh vegetables for personal consumption using organic gardening and sustainable water and energy. Natural fertilizers, companion planting, and water-efficient irrigation are eco-friendly agriculture practices. Off-grinders with limited arable land may use hydroponics or aquaponics for crop growth.

Off-grinders can employ permaculture, a holistic design method replicating natural ecosystems, to sustain land use. Fruit and nut trees, perennial herbs, and various plant species provide an autonomous environment with minimum inputs. Permaculture promotes biodiversity and ecological balance through human-environment symbiosis.

Livestock and small-scale animal husbandry promote off-grid self-sufficiency. Eggs, milk, and honey from chickens, goats, and bees give resources and connect people to the land. Off-gridders value sustainability and self-sufficiency, so they treat animals humanely.

They are living off-grid and demand constant learning and skill improvement. Off-graders learn renewable energy system maintenance, agriculture, and water conservation. Self-sufficiency requires problem-solving, adaptability, and learning new abilities.

Education goes beyond practical skills to include environmental care and ecological concepts. Off-gridders study local ecosystems, biodiversity, and sustainable land management. This understanding underpins responsible ecological behavior and decision-making.

Off-grid children study academics and life skills uniquely. Experiences in the off-grid lifestyle build a love of nature, ingenuity, and environmental responsibility. Off-grid households typically homeschool or choose alternative education options that fit their ideals and lifestyle.

Off-grid living requires careful consideration of upfront, ongoing, and long-term expenditures. Renewable energy systems, water infrastructure, and sustainable buildings may cost more than typical setups, but off-grinders value lower utility bills, self-sufficiency, and environmental effects.

Off-grid living demands budgeting and maintenance. Solar panels and batteries are durable and efficient, making off-grid setups economically viable. Lifetime considerations and proactive maintenance keep these systems reliable.

The economic equation demands a new view of consumption. Off-grinders prioritize needs above wants and use resources more sparingly. This approach promotes environmental awareness and responsible consumption in line with sustainability and self-sufficiency.

Off-grid living gives seclusion and independence but raises community and social issues. Off-grid living may be a choice for seclusion to escape urban life. However, seclusion can be rugged, especially for socially active people.

Off-grid communities, frequently founded by like-minded people committed to sustainability and self-sufficiency, foster belonging and shared ideals. These communities promote cooperation, knowledge sharing, and support. Off-grid life involves shared resources, community projects, and shared accountability.

If you want to live off the grid, you must evaluate your tolerance for isolation and independence. Lack of neighbors and service availability may necessitate more self-sufficiency and resilience. Off-grinders compensate for social isolation with excellent problem-solving abilities and a deep connection to the land.

In the off-grid lifestyle, technology is crucial to socializing. Digital communication, internet connectivity, and online communities save off-grinders from losing touch with friends, family, and networks. Off-grid residents must weigh the benefits of internet connectivity against their desire for a more natural lifestyle.

Sustainable Practices in Off-Grid Living

Adopting an off-grid existence demonstrates a strong bond with nature and a dedication to self-sufficiency. The quest for sustainability, a comprehensive strategy that goes beyond energy considerations to include water management, moral building methods, and the development of peaceful coexistence with the environment, is at the heart of this way of life. This section examines the many facets of off-grid living's sustainable practices, digging into the ideas supporting energy independence, water conservation, morale building, and developing an environmentally conscious and self-sufficient lifestyle.

Off-grid life is characterized by its pursuit of energy autonomy, attained by employing renewable resources. A vital component of this effort is solar power, which uses photovoltaic panels to turn sunshine into electricity. When installed on rooftops or ground, these panels offer a sustainable and environmentally friendly substitute for traditional power sources. They help lessen the environmental effects of conventional energy generation and reduce reliance on non-renewable resources.

In off-grid systems, wind energy supplements solar power by using wind kinetic energy to power small-scale wind turbines that produce electricity. Off-grid power systems benefit from an extra layer of robustness due to these turbines' strategic placement, which considers wind patterns and possible noise concerns. In addition, micro-hydro systems, which utilize the force of flowing water, offer an additional means of producing sustainable energy and demonstrate how off-grid living may be adapted to various geographic settings.

To achieve off-grid sustainability, energy storage is essential for bridging the intermittent nature of renewable resources. Lithium-ion and lead-acid batteries are advanced technologies that store extra energy when renewable resources are scarce. Combining energy-saving equipment with mindful consumption habits optimizes scarce resources, promoting a robust and sustainable energy system.

Water is a basic human need, and it becomes even more critical when living off the grid and there is no access to traditional municipal water sources. Rainwater collection stands out as a crucial tactic in sustainable water management. Off-grinders produce a decentralized, environmentally sustainable water supply by directing rainfall into storage tanks via rooftop collecting devices. Harvested rainwater is guaranteed to fulfill quality standards for domestic applications through filtration and purification techniques.

A well is another dependable water supply for off-grid requirements, which draws water from underground aquifers. Careful extraction techniques, routine water quality monitoring, and responsible, well-maintenance guarantee the long-term viability of this essential resource. To lessen the environmental effect of water extraction, off-grinders frequently prioritize energy-efficient pump systems that run on renewable energy.

Greywater systems play a significant role in water saving because they are made to treat and repurpose domestic wastewater. Greywater, produced during routine household chores like washing and showering, is treated before being used for irrigation or other non-potable purposes. In addition to reducing water waste, this closed-loop strategy supports the ideas of a sustainable and circular water economy.

Using water-efficient fixtures and appliances reinforces the off-grid lifestyle's emphasis on water conservation. Water-saving toilets, low-flow faucets, and well-planned irrigation techniques guarantee that every drop is used wisely. In addition to supporting environmental sustainability, prudent water resource management strengthens off-grid communities' ability to withstand the adverse effects of water scarcity.

Ethical and sustainable construction methods greatly influence the physical manifestation of off-grid life. An essential component of environmentally friendly architecture is passive solar design, which controls lighting and temperature by utilizing the sun's energy. Natural heating and cooling are facilitated by orienting structures to optimize sunshine exposure in the winter and to provide shade in the summer. Energy efficiency is promoted by large windows that are positioned to maximize sunshine and minimize the need for artificial lighting.

When building ethically, the selection of building materials is crucial. Reclaimed wood, recycled steel, and timber from sustainably managed forests are locally sourced and renewable materials that adhere to responsible resource use. Off-grid homes benefit from natural insulating materials like straw bales, recycled denim, and sheep's wool since they increase energy efficiency.

Using materials with strong heat retention capacities, known as thermal mass, is crucial to ethical construction. During the day, materials like rammed earth or adobe absorb and hold heat, releasing it gradually at night when it gets colder. By adding thermal mass, indoor climates can be made more consistent and comfortable without needing to use active heating or cooling systems as much.

Recycling and waste minimization techniques are essential components of ethical building. In off-grid buildings, reducing construction waste, repurposing materials, and using a cradle-to-cradle methodology supports a circular economy. Building with ethics demonstrates a dedication to minimizing environmental impact and designing homes that blend in with their surroundings.

Living off the grid frequently entails a dedication to sustainable farming and food production, focusing on permaculture, organic methods, and decreased dependency on outside food sources. With the help of sustainable energy and water sources, off-grinders can cultivate fresh vegetables through organic gardening. Eco-friendly agricultural practices include using natural fertilizers, companion planting, and water-efficient irrigation systems.

Off-grinders might employ permaculture, a holistic design philosophy influenced by natural ecosystems, to guide sustainable land usage. Permaculture integrates perennial herbs, various plant species, and fruit and nut trees to produce self-sustaining ecosystems with little help from outside sources. To promote biodiversity and ecological balance, the permaculture ideology strongly emphasizes the symbiotic interaction between humans and the environment.

One aspect of off-grid life that supports the self-sufficiency paradigm is livestock and small-scale animal husbandry. In addition to providing valuable resources, keeping chickens for eggs, goats for milk, and bees for honey helps people maintain a connection to the land. Off-gridders' emphasis on morality and the humane treatment of animals is frequently a priority, which fits with their dedication to sustainability and independence.

Off-grid waste management is a component of sustainable practices, emphasizing a comprehensive strategy that reduces impact and promotes a circular economy. Because there is less capacity for waste disposal in distant areas, off-gridders prioritize reducing waste output from the beginning.

In off-grid settings, composting is a vital part of waste management. Off-grinders use composting toilets to convert human waste into nutrient-rich compost, replenishing the soil with crucial organic matter. The cycle is completed by composting garden and kitchen waste, which establishes a closed-loop system that improves the soil fertility of the off-grid farmhouse.

Another essential component of trash management in off-grid living is recycling. Off-grinders choose products with minimum packaging and consciously reduce single-use packaging. Materials such as glass, metal, and some plastics are segregated for recycling, limiting the ecological impact of garbage disposal. Innovative upcycling and repurposing projects help reduce waste in some off-grid communities.

In addition to recycling and composting, off-gridders consider proper disposal of waste. Batteries and electronics are hazardous goods that must be handled carefully to avoid contaminating the environment. The off-grid lifestyle is defined by its broader values of self-sufficiency and environmental sensitivity, which align with responsible waste management.

Case Studies of Successful Off-Grid Communities

The appeal of living off the grid has drawn in people who want sustainability, independence, and a peaceful coexistence with the natural world. Off-grid communities represent the cooperative pursuit of resilience, environmental conscience, and self-sufficiency beyond individual pursuits. This section examines case studies of prosperous off-grid communities worldwide, examining the many strategies, difficulties encountered, and insights gained from these trailblazing sustainable living ventures.

Tinkers Bubble is a live example of the transformational potential of off-grid life centered around a community tucked away in the undulating hills of Somerset, England. This 40-acre intentional community was founded in 1994, and its members have made a deliberate decision to live a sustainable and cooperative existence. The community is self-sufficient through organic farming and permaculture techniques, and it depends on sustainable energy sources like solar and wind power to generate electricity.

Tinkers Bubble is characterized by its dedication to minimalism and eco-friendly living. The residents inhabit the self-built, environmentally friendly homes made of natural materials and adhere to ethical construction standards. The community is committed to minimizing its influence on the environment and promotes a sense of shared responsibility through group decision-making.

Tinkers Bubble emphasizes how crucial communal harmony is to sustainable off-grid living. A dedication to environmental stewardship, frequent social events, and shared responsibilities have all contributed to the success and longevity of this off-grid community. The need to develop a strong feeling of community and shared purpose is highlighted in Tinkers Bubble lessons as a means of overcoming the difficulties associated with off-grid life.

Earthaven Ecovillage, tucked away in the verdant Appalachian Mountains of North Carolina, is a shining example of regenerative life and holistic sustainability. This intentional community, established in 1994, is spread across 329 acres and features a variety of ecosystems, including woodlands and meadows. A thriving community of people who share a common vision of regenerative and sustainable living may be found in Earthaven.

The community's dedication to permaculture, organic farming, and renewable energy has made a robust off-grid life possible. Earthaven uses micro-hydro devices and solar panels to harness the natural energy flows of the land to produce electricity. The village also recycles greywater and collects rainfall to reduce its environmental impact.

One of Earthaven's unique selling points is its emphasis on varied and collaborative business endeavors. Locals work in various fields, including education, ecotourism, artisan crafts, and organic farming. This mixed economic strategy enhances the community's economic resiliency and self-sufficiency.

The success of off-grid life can be seen in the overall vision of Earthaven Ecovillage. The amalgamation of environmental, social, and financial sustainability cultivates a resilient and flexible community. This case study highlights the necessity of an all-encompassing strategy that considers a community's long-term vitality through economic diversity and shared values, in addition to its immediate demands for water and electricity.

Located on Scotland's northeastern shore, the Findhorn Ecovillage has gained international recognition as a symbol of sustainable living methods combined with spiritual principles. Since its founding in the 1960s, the community—comprised of more than 500 people from various backgrounds—has expanded to embrace environmental stewardship and overall well-being.

Innovative ecological design and sustainable infrastructure are among Findhorn's most well-known features. The neighborhood uses energy-efficient technologies and natural materials in its buildings, embracing passive solar design. The ecovillage uses solar panels, biomass heating systems, and wind turbines to provide renewable energy. Additionally, organic agriculture, permaculture, and sustainable water management are highly valued practices at Findhorn.

Findhorn is distinct in that it emphasizes spiritual and consciousness-based concepts. The residents practice everyday spiritual rituals, and the community values the connection between ecological and human well-being. The secret to Findhorn's success is its ability to create a balanced and harmonious way of life by fusing spiritual principles with realistic sustainability.

The case study of Findhorn Ecovillage demonstrates how spiritual principles can be integrated with environmentally friendly methods. This community is a prime example of how a solid spiritual bond and a close relationship with the natural world can motivate and uphold an ecologically conscious lifestyle. The takeaways from Findhorn emphasize how crucial it is to include mindfulness and a sense of purpose in your off-grid lifestyle.

Tucked away in the isolated wilderness of the Strait of Georgia, off the eastern shore of Vancouver Island, Lasqueti Island is a live example of the opportunities and difficulties that come with living off the grid in a remote area. Lasqueti Island, home to about 400 people, has developed into a haven for people looking for independence and living in harmony with the natural world.

The primary sources of electricity generated on Lasqueti Island are micro-hydropower and solar energy. Residents of the island have adopted energy-conscious living habits and energy-efficient appliances as part of a thrifty and sustainable approach to energy consumption. The community emphasizes water conservation obtained from wells and rainwater gathering.

Lasqueti Island stands out for its conscious, low-impact lifestyle and dedication to autonomy. Many locals are homeowners who built their own homes from scratch using repurposed materials and a minimalist style. The island's population actively participates in local governance and decision-making processes despite providing meager public services.

The story of Lasqueti Island sheds light on the difficulties and benefits of off-grid life in isolated areas. The islanders have become remarkably self-sufficient but have also had to deal with the logistical challenges of getting goods and services. Lasqueti Island is a perfect example of how, despite geographical isolation, people must be flexible, inventive, and bonded by a strong sense of community.

Crystal Waters Permaculture Village, located in Queensland, Australia's outback, is a model for incorporating permaculture ideas into off-grid community living. More than 200 people call Crystal Waters, which was founded in the late 1980s, home. Permaculture is the community's guiding philosophy for sustainable living.

The community uses a holistic approach to land use, with a strong emphasis on energy efficiency, organic farming, and sustainable building techniques. Using a combination of solar panels, wind turbines, and micro-hydro systems, Crystal Waters produces electricity, demonstrating the adaptability of renewable energy sources in a subtropical environment. Residents use greywater systems and rainfall harvesting because conserving water is very important.

The structure of Crystal Waters, a regenerative community, is based on permaculture principles, incorporating wildlife habitats, food forests, and community gardens into the overall plan. The village promotes economic diversification through neighborhood companies, community-supported agriculture, and ecotourism programs.

Crystal Waters' dedication to using permaculture as a guiding design philosophy makes it successful. The adoption of regenerative practices by the community serves as an example of how permaculture concepts may be extended beyond agriculture to include other facets of community life. The case study highlights how careful planning is essential to building a robust and sustainable off-grid community.

CHAPTER VIII

Budgeting and Financial Considerations

Initial Investment and Cost Breakdown

Starting an off-grid lifestyle entails a significant paradigm change, including giving up traditional utilities and embracing self-sufficiency. Although the appeal of sustainability, energy independence, and a peaceful relationship with the natural world is strong, the financial implications of moving to an off-grid lifestyle should be carefully considered. This section delves into the subtleties of the necessary initial outlay and offers a comprehensive cost breakdown for individuals contemplating or having already ventured into off-grid living.

The choice to live off the grid is primarily influenced by the initial expenses related to building a self-sufficient infrastructure. The primary financial commitment is to renewable energy systems, essential to off-grid living. For off-grid households, solar power frequently acts as the primary source of electricity due to its accessibility and dependability. Most upfront costs comprise the first investments in energy storage devices like batteries, inverters, and solar panels.

Another feasible option is wind energy. However, this requires installing wind turbines, which raises the initial cost. The financial concerns include the expenditures related to choosing the right turbine size, ensuring it is installed correctly, and integrating it into the broader energy system. Micro-hydro systems require upfront spending on equipment, such as turbines and control systems, that are customized to the site's unique features to capture the energy from flowing water.

Off-grid living's financial sustainability depends on how long-lasting and effective renewable energy sources are. Investing in high-grade solar panels, inverters, and batteries is necessary to build a dependable energy system. The long-term advantages of decreased utility costs, energy independence, and a minor environmental effect outweigh the more considerable initial expenditure in the long run.

The funding commitment covers water infrastructure in addition to energy systems. Off-grid homes rely on wells, rainwater collection systems, or other dispersed water sources. The initial investment is financed mainly by the price of establishing water collection systems, storage tanks, filtration, and purification systems. The cost breakdown becomes more complex when considering greywater recycling, renewable energy-powered pump systems, and responsible well drilling.

Off-grid living is known for its ethical construction methods, which add further factors to the financial picture. Naturally occurring insulation, recycled steel, and locally sourced wood are sustainable and environmentally friendly building materials, but they can be more expensive initially. Design considerations and construction methods that increase the initial investment may be part of the concepts of passive solar design, which are essential to energy-efficient buildings.

Composting toilets are one of the environmentally-conscious features that off-grid homes with a low environmental impact may include, but they also have an associated cost. Permaculture-inspired landscaping, rain gardens, and waste management systems influence the total cost of using ethical construction methods.

Regarding water and waste management, the initial outlay for rainwater harvesting solutions, composting toilets, and greywater systems aligns with sustainability standards. These systems have particular equipment, installation, and continuing maintenance expenses, even if they promote resource efficiency and environmental stewardship.

Off-grid life frequently involves food production and agricultural pursuits; people grow organic gardens, engage in permaculture, and raise small animals. The initial costs of establishing these sustainable food production systems are associated with seeds, fencing, animal houses, soil amendments, and livestock procurement. Creating diversified ecosystems, adding fruit and nut trees, perennial herbs, and other regenerative practices, and investing in permaculture principles can all add to the initial outlay.

Off-grid living requires a lot of educational considerations because inhabitants frequently need to pick up various skills, from permaculture techniques to renewable energy system maintenance. The initial investment includes workshops, seminars, and instructional materials that create a learning environment where people can thrive in a self-sufficient lifestyle. This dedication to lifelong learning supports the community's long-term viability and is consistent with the off-grid lifestyle.

Off-grid living's economic calculation thoughtfully weighs up-front expenses against long-term sustainability and independence. Compared to standard setups connected to municipal utilities, the initial expenditure may be more significant, but in most cases, the long-term advantages outweigh the costs. The financial feasibility of living off the grid is influenced by fewer environmental impacts, more self-sufficiency, and decreased or eliminated utility expenses.

For off-grid living to be successful, careful financial planning and consideration of ongoing maintenance expenditures are essential. A critical factor in the economic calculation is the robustness and efficiency of renewable energy components, like solar panels and batteries. Warranty conditions, proactive maintenance schedules, and lifespan considerations guarantee these systems' long-term dependability.

The economic benefits of off-grid living also entail a change in viewpoint on spending habits. Off-grinders prioritize needs above wants and take a more deliberate and thrifty approach to resource usage. This way of thinking supports a way of life that emphasizes responsible consumerism and environmental awareness, consistent with sustainability and self-sufficiency.

It is critical to recognize that there is no one-size-fits-all initial outlay for off-grid living. Depending on several variables, including location, climate, resources available, and desired degree of self-sufficiency, the expenses can differ significantly. Logistical difficulties in remote areas might affect the price of material transportation and specialized service access. On the other hand, areas rich in renewable resources can reduce energy production costs.

Off-grinders frequently highlight the non-financial advantages of their lifestyle choice, including improved health, a stronger bond with the natural world, and the fulfillment of leaving a minor environmental mark. Living in harmony with the environment, becoming more self-sufficient, and building a feeling of community have inherent values that frequently outweigh short-term financial concerns.

An additional layer to the financial environment of off-grid life is the consideration of laws and regulations. Regional variations in zoning laws, building standards, and environmental regulations affect the viability and cost of off-grid living. While some regions may encourage off-grid and sustainable building techniques, others can limit non-traditional building techniques or need a connection to city services.

Off-grinders must be proactive and knowledgeable to navigate these legal problems. Planning and carrying out off-grid projects requires knowing zoning laws, interacting with local officials, and following construction requirements. The whole cost of financial planning for off-grid life should account for the expenses related to legal and regulatory compliance.

Off-grid living may seem like a big initial commitment, but it's essential to consider it a long-term, strategic decision. The economics of off-grid life consider resilience, self-sufficiency, and environmental stewardship in addition to a straightforward cost-benefit analysis. Financial obstacles to off-grid living are anticipated to change as economies of scale increase, technology develops, and public attitudes toward sustainability change, opening up this lifestyle option to a broader range of people.

In summary, the upfront costs and cost breakdown of off-grid living encompass many intricately intertwined elements, from trash disposal and water management to ethical building practices and renewable energy systems. While significant up-front fees may exist, the long-term, intrinsic, and economic advantages often outweigh the cost. As off-grid living gains popularity as a feasible and sustainable lifestyle option, navigating the economic terrain of this revolutionary trip will require careful preparation, constant learning, and flexibility.

Government Incentives and Rebates

Government subsidies and incentives are crucial in determining how off-grid initiatives develop in the dynamic world of sustainable living. As society faces the urgent concerns of environmental degradation and climate change, governments everywhere realize how important it is to support eco-friendly practices, such as using off-grid technologies. This section explores how policy measures might accelerate the shift toward a more resilient and ecologically conscious future by delving into the wide range of government incentives and rebates intended to encourage individuals and communities to embrace sustainable, off-grid living.

The government is incentivizing the adoption of renewable energy, which is a leading support for living off the grid. Businesses and people investing in solar panels, wind turbines, and other renewable energy sources can receive financial incentives from many nations. These financial aids, which typically take the shape of tax credits, grants, or refunds, significantly reduce the up-front expenses related to the construction of renewable energy infrastructure. Governments hope to encourage wider adoption of sustainable energy technologies, increasing energy independence and reducing dependency on conventional power sources by lowering financial barriers.

The Federal Investment Tax Credit (ITC) in the US is one prominent example; it offers a sizeable tax credit for residential and commercial solar installations. The country's solar power industry has dramatically increased thanks to the Investment Tax Credit (ITC), which encourages utility-scale projects, enterprises, and households to adopt solar energy. Similar initiatives are in place in several nations, customizing incentives to meet the unique requirements and difficulties of each nation's energy environment.

Governments are beginning to understand how critical it is to assist localized energy production in isolated and off-grid locations. Where it is not financially feasible to expand the conventional grid infrastructure, governments might offer further incentives to encourage the use of off-grid energy sources. In addition to addressing energy poverty, this strategy builds resilient microgrids that are self-sufficient and resilient enough to endure grid failures.

Aside from energy, a vital component of off-grid life is water management, and governments have put incentives in place to promote sustainable water practices. Government subsidies or grants are frequently available for rainwater collecting, greywater recycling, and water-saving initiatives. These programs not only ease the burden on municipal water supplies but also provide communities and individuals the ability to manage their own water requirements in off-grid environments.

In the context of environmentally conscious building and ethical construction, governments realize how important it is to incentivize sustainable building methods. Utilizing eco-friendly building techniques, passive solar design, and energy-efficient materials may qualify for tax credits or rebates. This encourages sustainable living and helps lower the carbon footprints left by conventional building methods.

Local governments are frequently adopting zoning and building code modifications to support off-grid living. In some circumstances, legislation may be amended to enable alternative energy systems, water catchment systems, and composting toilets. These adjustments show a growing awareness of sustainable living methods and a readiness to assist locals who lead different lives.

Governments are promoting sustainable living with a comprehensive strategy, further demonstrated by recycling and trash reduction incentives. Financial incentives may be available to people and communities who adopt recycling programs, composting systems, and other waste reduction measures. Governments help create closed-loop systems with minimal environmental impact by promoting appropriate waste management.

Government incentive programs pay attention to the agricultural aspect of off-grid life. Agroforestry, permaculture, and organic farming may be supported by subsidies, grants, or technical assistance from nations prioritizing sustainable agriculture. These financial incentives support the growth of resilient, regenerative agricultural systems consistent with off-grid living ideals.

Success in off-grid living requires education and training,

and governments understand that developing skills and expertise in sustainable practices is essential. Government efforts may support or sponsor workshops, training programs, and educational materials centered on permaculture, renewable energy, and ethical construction. Investing in human capital can create a more knowledgeable and capable populace that can successfully navigate the opportunities and difficulties associated with off-grid living.

Remembering that government rebates and incentives

vary by area and nation is crucial. These programs' designs and scopes differ according to local political conditions, resources, and priorities. Regarding incentive programs, national governments may take the lead, but in other situations, regional or local government agencies are more important.

Government incentive programs must be well-designed,

communicated, and flexible enough to change with the times to be successful. Governments must balance long-term planning, fiscal prudence, and the desire to promote sustainable behaviors. Policymakers can evaluate incentive programs regularly to see how they are doing, pinpoint areas for development, and ensure they align with larger social and environmental objectives.

Even if government grants and incentives offer

substantial assistance for living off the grid, there are obstacles and restrictions. These programs may not always be readily available or easily accessible, and bureaucratic procedures occasionally make quick installation difficult. Furthermore, political will must be demonstrated by continuing and growing incentive programs over time to provide a stable and encouraging climate for off-grid projects.

To sum up, government rebates and incentives significantly influence the off-grid living environment by offering financial and legal support to people and communities who want to pursue sustainable lifestyles. Adopting renewable energy, sustainable agriculture, water management, and ethical building are just a few of the areas where governments worldwide realize the complexity of off-grid life and adjust incentive schemes accordingly. These government programs act as catalysts for the shift to more resilient, environmentally friendly, and self-sufficient modes of living as the world struggles to find lasting answers.

Long-Term Cost Savings

The decision to embrace an off-grid lifestyle is a commitment to environmental sustainability and an intelligent move toward long-term financial savings. Despite the initial high cost of sustainable infrastructure, renewable energy systems, and self-sufficient lifestyles, off-grid living's economic resilience emerges gradually. This section examines how living off the grid lowers long-term costs, from energy independence and lower utility bills to more sustainable food production, effective resource management, and less dependency on outside services.

One of the main factors contributing to long-term savings when living off the grid is the decrease or do away with typical utility expenses. Monthly costs for energy, water, and sewer services are recurring for conventional homes connected to municipal power networks and water suppliers. On the other hand, people who live off the grid and use renewable energy sources, such as wind turbines or solar panels, see a significant decrease in their electricity expenses. Even while these systems may require a more effective initial expenditure, over time, they save more and more, giving users predictable and stable financial results.

Living off the grid promotes energy independence, protecting people and communities against utility price fluctuations and possible outages in centralized energy infrastructure. Fuel price swings and geopolitical unrest can affect electricity pricing in the global energy environment. Off-grid homeowners protect themselves from outside forces that can result in erratic and escalating energy costs by producing electricity locally using renewable resources.

Furthermore, improvements in energy storage technologies—like effective battery systems—improve off-grid systems' capacity to retain extra energy produced during sunny or windy spells. When there is a lack of renewable resource availability, this stored energy becomes a vital resource that guarantees a steady and dependable power supply. Off-grid energy storage can save costs because it doesn't require backup generators or grid-connected support, thanks to its flexibility and durability.

Off-grid living offers significant long-term savings in the field of water management. Monthly invoices for municipal water services include sewage fees, consumption-based billing, and increasing costs during dry spells or water shortages. Water-related expenses are significantly reduced for off-grid residents who use wells, rainwater collection, or dispersed water sources. Rainwater becomes a free and sustainable resource for home use, livestock, and irrigation when gathered and stored in tanks.

Reusing greywater is a popular off-grid method that increases water efficiency even more. Greywater systems treat and repurpose spent water from activities like laundry and showering for non-potable uses like landscape irrigation rather than letting it go to waste. By lowering dependency on municipal water sources and wastewater disposal costs, this closed-loop method saves water use and contributes to long-term savings.

Off-grid living frequently entails ethical construction methods, which save money in the short and long terms. The longevity and energy efficiency of locally sourced and sustainable materials lead to lower maintenance and operating costs over time, even though their use may initially result in higher expenditures. Passive solar architecture and well- insulated buildings use less energy for heating and cooling, which reduces utility costs and improves comfort without requiring constant reliance on active temperature control systems.

Composting toilets are another common feature found in off-grid homes; these units support environmental sustainability and save money over time. Conventional sewage systems require periodic upkeep, possible repairs, and connection fees to city services. Contrarily, composting toilets break down human waste on the spot, creating nutrient-rich compost that can be added to the soil. In addition to lessening the environmental impact, this decentralized method does away with the requirement for ongoing sewage fees and services.

Sustainable food production methods enable long-term cost reductions in the agricultural aspect of off-grid living. To produce fresh vegetables and sources of protein, off-grid communities frequently practice organic gardening, permaculture, and small-scale animal husbandry. Off-grinders decrease their dependency on grocery shops and cut costs on fruit, meat, and dairy goods by growing their food.

Regenerative and sustainable agriculture, the cornerstones of permaculture, further bolsters off-grid food production's long-term economic viability. Permaculture reduces the demand for fertilizers, insecticides, and water resources by establishing self-sustaining ecosystems that require little outside inputs. This strategy supports biodiversity and soil health while eventually saving money.

Living off the grid fosters a deliberate and thrifty approach to resource usage, discouraging wasteful spending and promoting long-term savings. Off-grid individuals and groups frequently embrace minimalist lives, putting needs before wants and purposefully abstaining from unnecessary expenditures. This way of thinking is consistent with the values of sustainability and thrift, which results in less money spent on unnecessary products and services.

Off-grid living has long-term economic reductions that also apply to trash management techniques. In addition to reducing environmental effects, composting systems, recycling programs, and appropriate waste disposal also boost economic efficiency. By building closed-loop systems that recycle organic waste into compost and repurpose recyclable materials, off-grid communities decrease the need for external waste disposal services, landfills, and associated fees.

Continuous cost savings in off-grid energy systems are facilitated by economies of scale and technological advancements. Over time, advances in the longevity and efficiency of energy storage technologies, wind turbines, and solar panels have increased the overall return on investment for off-grid installations. With the rising adoption of these technologies, costs should continue to decline, hence boosting the accessibility and viability of off-grid living.

Government incentives and rebates encourage the adoption of sustainable practices and renewable energy, which also contribute to the long-term financial viability of off-grid living. Governments encourage people and communities to use off-grid technologies by offering financial incentives and covering early costs. This creates a mre level playing field and promotes widespread adoption.

Although living off the grid can result in significant long-term cost savings, it is essential to recognize that the economic picture changes depending on several circumstances, including temperature, location, and desired level of self-sufficiency. Logistical difficulties in remote areas might affect the price of material transportation and specialized service access. On the other hand, areas rich in renewable resources can offer chances for higher energy production cost savings.

In summary, the long-term financial benefits of living off the grid are evident in several areas of daily life, such as lower electricity costs and sustainability.Water management, moral building methods, and economical use of resources. Off-grid living's economic resilience stems from its capacity to reduce dependency on outside services, utilize renewable resources, and promote resource conservation consciousness. Off-grid living is expected to become more financially viable as technology develops, economies of scale increase, and public attitudes toward sustainability change. This will open the option to a broader range of people who value environmental sustainability and sound financial management.

CHAPTER IX

Future Trends in Off-Grid Solar Power

Advancements in Solar Technology

Recent years have seen remarkable developments in solar technology, which have entirely changed the renewable energy environment and are essential to the world's transition to sustainability. With advancements in energy storage, creative applications, and photovoltaic efficiency, the development of solar technology offers hope for a more sustainable and productive energy future. This section examines the main developments in solar technology and their effects on energy production, the environment, and the general uptake of solar energy.

The ongoing enhancement of photovoltaic (PV) cell efficiency is among the most notable developments in solar technology. PV cells, which transform sunlight into electricity, are the central component of solar panels. To maximize energy output, scientists and engineers have worked to improve these cells' efficiency over time. Innovations like bifacial solar cells and passivated emitter rear contact (PERC) technology have increased the efficiency of traditional silicon-based solar cells, which were once the industry standard.

For example, the back-surface passivation method used in PERC technology lowers electron recombination and facilitates more effective energy conversion. Bifacial solar cells, which can collect light from the front and the back of the sun, increase the energy produced using reflected light from nearby surfaces. Because of these improvements in PV efficiency, more energy is produced, which increases the viability and competitiveness of solar power as an electrical source.

Thin-film solar technology is another innovative field in pursuing higher efficiency and lower manufacturing costs. Thin-film solar cells, which are lighter and thinner than conventional silicon-based cells, use layers of semiconductor materials like copper indium gallium selenide (CIGS) or cadmium telluride (CdTe). With its adaptability, this technology finds use in various contexts, such as portable solar gadgets and building-integrated photovoltaics (BIPV).

When mitigating the intermittent nature of solar power generation, combining solar technology and energy storage devices has changed the game beyond efficiency gains. Lithium-ion batteries are one type of energy storage option that allows for storing extra energy produced during peak solar hours for use during low solar irradiance or high demand. One of the long-standing obstacles to integrating renewable energy sources is overcome by combining solar power and improved energy storage, which improves grid stability and provides a more dependable and constant power supply.

Energy capture has been further enhanced through solar tracking devices that ensure that solar panels align with the sun's position throughout the day. With single- and dual-axis solar trackers, solar panels' orientation may be dynamically changed to maximize sun exposure and boost energy production. Through the optimization of the angle of incidence and the mitigation of shadowing effects, these tracking systems—which are frequently used in utility-scale solar installations—improve the overall efficiency of solar power plants.

A new frontier in solar innovation is the development of organic photovoltaic (OPV) technology, which holds promise for flexible, inexpensive, and lightweight solar panels. Organic molecules serve as the active layer in OPV cells, enabling charge production and light absorption. Even though OPV technology now faces efficiency issues compared to conventional solar cells, continuous research and development activities are intended to enhance the technology's performance. Because organic materials are flexible and adaptable, OPV cells can be used for non-traditional applications like solar-integrated clothes and portable, lightweight solar systems.

Concentrated solar power (CSP) systems have seen developments in solar thermal technology that improve their effectiveness and practicality. CSP systems use mirrors or lenses to focus sunlight onto a small area to produce electricity or other industrial processes. Utilizing cutting-edge heat absorption and storage materials, including molten salts, improves system performance overall and enables CSP to create energy longer than the day.

Transparent solar cells are a revolutionary development in urban planning and architectural integration. Solar windows, or transparent solar cells, can be installed in windows or building facades without blocking natural light. These translucent solar panels provide an aesthetically beautiful and energy-efficient option for sustainable building design by capturing sunlight while letting visible light through. Incorporating solar technology into architectural components can completely transform how energy is used in urban settings.

The field of materials science has been crucial in propelling progress in solar technology by investigating new materials that offer improved sustainability, durability, and efficiency. For example, perovskite solar cells have become a viable substitute for conventional silicon-based cells. Perovskite materials may be processed economically and have outstanding light-absorbing qualities. The main goals of current research are to solve stability problems and increase the output of perovskite solar cells for industrial usage.

The idea of solar paint, sometimes known as photovoltaic paint, is a creative way to apply solar technology to commonplace surfaces. Scholars are investigating the creation of paints or coatings that incorporate compounds that absorb light, enabling surfaces such as roofs or walls to produce electricity. The prospective uses of solar paint in infrastructure, architecture, and even transportation, albeit still in the early phases of development, hold promise for extending the reach of solar energy generation.

Solar technology innovations have merged with the Internet of Things (IoT) and smart grid technologies to provide energy systems that are more responsive and intelligent. Bright solar panels with sensors and communication capabilities can maximize energy production based on real-time data, weather, and grid demand. This degree of connectedness makes it possible to manage energy more effectively, waste less, and strengthen grid resilience overall.

Recognizing solar technology's transformative potential, governments and corporations invest more in research, development, and large-scale implementation. The solar sector is growing due to incentive programs, subsidies, and regulatory frameworks that encourage the use of solar energy. The global adoption of solar power has been further expedited by the falling costs of solar components, fueled by economies of scale and technological advancements.

Developments in solar technology have ramifications that go beyond producing electricity and tackle more general environmental issues. By shifting to solar energy, we can lessen our reliance on fossil fuels and reduce the adverse effects of air pollution and greenhouse gas emissions on the environment. Decentralized solar energy generation fosters resilience against external shocks and contributes to a more robust and distributed energy infrastructure, enabling communities to become energy-independent.

Although solar technology is advancing and has immense potential, there are still obstacles to its general implementation. Sunlight intermittency, the requirement for practical energy storage systems, and the environmental effects of producing and discarding solar components are some issues that demand constant research and development. To guarantee that the benefits of solar developments are inclusive and reach marginalized areas, it is also necessary to prioritize the equitable distribution of solar benefits and accessibility to clean energy technologies.

Emerging Trends in Energy Storage

Energy storage plays a more and more critical role as the world works toward a robust and sustainable energy future. Energy storage technologies are essential to mitigate the intermittent nature of renewable energy sources, maintain grid stability, and facilitate the shift to a low-carbon energy future. This section covers the developing trends in energy storage, evaluating breakthroughs in battery technology, the advent of novel storage solutions, and the transformative impact these trends have on the global energy paradigm.

The development of battery technologies is one of the most critical trends in energy storage. Long regarded as the industry standard, lithium-ion batteries have undergone constant growth, leading to higher energy density, longer cycle lives, and reduced prices. The economies of scale for lithium-ion batteries have been fueled by the increasing use of electric vehicles (EVs), resulting in a sharp decline in price and increased accessibility for various energy storage applications.

Researchers and industry participants are looking into alternative battery chemistries that have benefits over lithium-ion, including those related to performance, sustainability, and safety. For example, solid-state batteries could improve safety and energy density by substituting a solid electrolyte for the liquid electrolyte found in conventional batteries. With the ability to overcome issues related to thermal runaway and enable larger storage capacity, solid-state batteries have the potential to transform both consumer devices and grid-scale energy storage completely.

Another exciting development in energy storage technology is the use of flow batteries. Flow batteries store energy in liquid electrolytes housed in external tanks, unlike conventional batteries that store energy in electrodes. Because of their design, flow batteries may expand their energy capacity and power independently, making them appropriate for large-scale grid applications. Vanadium redox flow batteries, in particular, have been popular due to their robustness, scalability, and capacity for long-term storage, which helps mitigate the unpredictability of renewable energy production.

Hybrid energy storage systems are becoming increasingly popular since they combine several storage technologies for mutually beneficial effects. The system's performance is improved when lithium-ion batteries are combined with complementing technologies like flywheels or supercapacitors. A dynamic and adaptable energy storage solution is produced when supercapacitors' quick response times and batteries' high energy densities are combined. By addressing the shortcomings of separate technologies, these hybrid systems provide more flexibility and dependability in managing energy storage systems.

The rise in popularity of electric vehicles has led to the electrification of transportation, facilitating the development of vehicle-to-grid (V2G) technology. With the help of vehicle-to-grid (V2G) technology, electric cars (EVs) can feed extra energy back into the grid and consume it. Through this reciprocal relationship, EVs become mobile energy storage devices that optimize the use of renewable energy sources, support the grid during periods of high demand, and lessen the burden on the electrical infrastructure.

In decentralized power management, distributed energy storage systems at the household or community level constitute a revolutionary movement. Home energy storage systems, frequently combined with rooftop solar panels, enable users to store extra energy produced during the day for use in high demand or low solar irradiance. These technologies help to democratize energy generation and consumption while improving energy resiliency and lowering dependency on centralized grid infrastructure.

Energy storage systems rapidly incorporate artificial intelligence (AI) and machine learning to improve predictive capacities and maximize efficiency. AI algorithms determine in real-time whether to charge or discharge storage systems based on historical data, weather patterns, and trends in energy demand. By enabling adaptive reactions to shift grid conditions and enhancing the overall stability of renewable energy integration, this intelligent control increases the efficiency and dependability of energy storage.

Trends in energy storage are being influenced by the emergence of green hydrogen as an energy carrier, especially in areas with a surplus of renewable resources. Hydrogen is produced through electrolysis, fueled by renewable electricity, and results in a clean, adaptable energy storage medium. When direct electrification or other impractical storage methods are used, green hydrogen can be transferred and stored to provide a long-term energy storage alternative. The decarbonization of industries, transportation, and heating is aided by adding green hydrogen to the energy mix.

Large-scale, long-duration energy storage is becoming increasingly possible with the help of gravity-based energy storage devices, such as kinetic and gravitational potential energy storage. These devices store energy by lifting large objects or producing rotational momentum using the laws of physics. These masses are reduced, or the rotating energy is transformed back into electricity when energy is required. Gravity-based storage has the benefit of scalability and possible cost-effectiveness for long-term energy storage.

Communal energy storage is becoming increasingly popular, encouraging cooperation and resource sharing among neighbors. The implementation of shared storage infrastructure for the benefit of numerous homes or businesses is known as community energy storage. This strategy allows communities to share resources, use energy more efficiently, and participate in grid services. In addition to increasing local energy resilience, community energy storage fosters sustainability and a sense of shared accountability.

Research is being done on blockchain technology to improve energy storage systems' efficiency, security, and transparency. Peer-to-peer energy trading is made more accessible, and the decentralized and secure transactions made possible by blockchain technology encourage the contribution of stored energy to the grid. Blockchain-enabled energy storage devices can participate in grid balancing services, facilitating a smoother integration of renewable energy supplies and promoting a decentralized energy market.

An essential factor in the development of energy storage systems is environmental sustainability. The sector is being shaped by measures to recycle, move toward more sustainable materials, and lessen the ecological effect of manufacturing processes. To guarantee that energy storage solutions comply with the tenets of a circular economy, producers and researchers are looking into substitute materials for those that are scarce or have adverse environmental effects.

Governmental incentives and regulatory frameworks significantly shape energy storage technology deployment. Governments everywhere realize how vital energy storage is to meeting renewable energy targets and improving system stability. Encouraging policies, such as tax breaks, grants, and expedited approval procedures, stimulate investments in energy storage systems and foster an atmosphere favorable to innovation and implementation.

Challenges and considerations accompany the encouraging advancements in energy storage. Among the many complicated challenges that need to be addressed are the effects of manufacturing and disposing of storage technology on the environment, the need for standardized regulatory frameworks, and the integration of various storage options into the current energy infrastructures. For energy storage to evolve sustainably, finding a balance between technological innovation, economic feasibility, and environmental stewardship is still a significant challenge.

In summary, new developments in energy storage portend a revolution in the energy world. The way societies generate, store, and consume energy is changing due to advancements in battery technology, the proliferation of innovative storage options, and the integration of intelligent control systems. Energy storage becomes essential for building a resilient, sustainable, decentralized energy future as nations strive to meet aggressive renewable energy objectives and tackle climate change. To fully realize the promise of these trends and expedite the shift to a cleaner and more sustainable energy paradigm, companies, governments, and researchers must continue to collaborate.

CONCLUSION

"Sunlit Solutions: A Practical Handbook for Off-Grid Solar Power" concludes with a thorough introduction to the liberating world of off-grid solar living. This guidebook comprehensively examines the various facets of using solar energy for off-grid, sustainable living, going beyond the typical bounds of solar literature. The book guides readers through the complex world of solar power, from the fundamentals of solar energy and identifying critical loads to the nuances of solar panel installation and battery storage solutions, all with an unrelenting commitment to clarity and pragmatism.

"Sunlit Solutions" excels not only in explaining technical aspects but also in emphasizing practical applications. The handbook provides valuable insights into seasonal changes, energy storage, and solar panel selection. This ensures that theoretical knowledge is effectively used in real-world scenarios. Including case studies and forward-thinking research of new trends lends a dynamic component to the book, catering to both novices and seasoned practitioners.

In addition to providing information, "Sunlit Solutions" fosters an attitude that values sustainability, resiliency, and independence. It helps readers develop a deeper comprehension of the subject matter and inspires them to adopt an off-grid existence intentionally. Utilizing solar electricity requires a dedication to holistic living, reflected in integrating factors like energy conservation, smart appliances, and ecological behaviors.

Through the chapters of "Sunlit Solutions," readers are taken on a trip beyond traditional energy assumptions. The book acts as a beacon, pointing people and communities toward a time when solar energy will be a way of life rather than merely a new technology. In a world where sustainable solutions are needed to address environmental issues, "Sunlit Solutions" is a relevant and helpful resource that gives readers the confidence and purpose to traverse off-grid solar power's complexities successfully. This guidebook shows a bright future where sustainable energy solutions enable people to live freely and ethically, regardless of whether they are starting off-grid for the first time or want to gain more knowledge.

Thank you for buying and reading/ listening to our book. If you found this book useful/ helpful please take a few minutes and leave a review on the platform where you purchased our book. Your feedback matters greatly to us.